Upstream Engineering and Flow Assurance Forum 2019

Held at the 2019 AIChE Annual Meeting

Orlando, Florida, USA
10-15 November 2019

ISBN: 978-1-7138-0546-5

Printed from e-media with permission by:

Curran Associates, Inc.
57 Morehouse Lane
Red Hook, NY 12571

AIChE
120 Wall Street, FL 23
New York, NY 10005-4020

Phone: (800) 242-4363
Fax: (203) 775-5177

www.aiche.org

Additional copies of this publication are available from:

Curran Associates, Inc.
57 Morehouse Lane
Red Hook, NY 12571 USA
Phone: 845-758-0400
Fax: 845-758-2633
Email: curran@proceedings.com
Web: www.proceedings.com

TABLE OF CONTENTS

Experimental Characterization of Surfactant Application in Gas Wells and in Presence of Water and Condensates

Amita Kosambi[1], Hamidreza Karami[1] and Sepideh Razavi[2], (1) MPGE, University of Oklahoma, Norman, (2) Chemical Engineering, University of Oklahoma, Norman, OK

Abstract

Liquid loading or inability of gas to carry the accumulated liquids along the wellbore shortens the life of Natural Gas wells, costing the operators in revenue. One of the common gas-well deliquification techniques is surfactant (foamer) injection or assisted foam lift. This method is proven very beneficial especially at low-pressure gas reserves, where the required critical gas velocity to avoid liquid loading gets reduced to around 20% of its original value. However, it is known that the presence of condensed hydrocarbons lowers the surfactant's efficiency in foam lift operations. Gas wells operating at higher water cut are good candidate for foam lift as presence of water enhances the foaming and helps the unloading. This study is intended to experimentally evaluate the performances of various surfactant types in presence of both water and hydrocarbon condensates with regards to unloading the liquids.

The purpose of this study is to relate the fundamentals of surfactant characterization with foam lift efficiency. Five different types surfactants are selected for the study. Experimental work includes static and dynamic surface tension measurements and liquid column unloading tests with varying surfactants and concentrations. The unloading tests are completed at the laboratory scale with changing liquid column water cuts and a fixed air injection rate to generate foam. The results are recorded in terms of unloaded liquid, foam quality, liquid drainage rate and foam half-life. A noticeable drop can be observed in liquid unloading as the water cut is reduced, showing the diminishing foam lift efficiency. However, the significance of this drop is different for different surfactant types.

Introduction

Over the life of gas production wells as the pressure is declining, they face the issue of liquid loading. Gas wells commonly produce water and hydrocarbon condensates along with the gas. Due to the reduced reservoir pressure, the gas velocity becomes insufficient to carry the liquids to the surface. As a result, the well starts functioning with interruption as the well bottomhole is loaded with liquids. Along with the low gas production, operators lose the hydrocarbon condensates, which are valuable and have a higher calorific value than gas. Eventually the well might die if not treated in time. Operators need to look at various ways to resolve this issue.

The recent outlook of low oil and gas prices encourages operators to produce from existing mature reservoirs instead of exploring new reserves. This means that innovative technologies are needed to artificially lift and improve the production of existing declined gas reserves. Foam lift is one of the verified well deliquification techniques for natural gas wells. In this method, a surfactant (foamer) is injected downhole continuously or in batches, and the liquid is unloaded within the foam structure. Usage of foamer is cost efficient as it doesn't require any external energy supply. The challenging part with foam lift is to maintain the performance of the surfactants in presence of gas condensates. This study aims to experimentally evaluate the performances of various surfactants in presence of both water and condensates. Firstly, a review of the available literature is presented.

One of the most comprehensive works on foam lift is completed by Ajani (2014), an experimental and modeling study of the effects of surfactants on liquid loading in vertical gas wells. The surfactants investigated were five types of commercial surfactants characterized into anionic, amphoteric and cationic classes. The objective was to study the effect of surfactants in terms of liquid unloading in 2-inch and 4-inch ID, 40 feet vertical pipes. Also a model was developed utilizing the results from bench top tests, predicting pressure gradient of foam flow. The observations were recorded for liquid holdup, pressure gradient, foam flow regime, and the required critical superficial gas velocity. In terms of liquid unloading, the results showed the usage of surfactant has delayed the liquid loading in gas wells. Increasing the surfactant concentration beyond the CMC did not prove beneficial for liquid unloading. Regarding different surfactant types, he observed the highest unloading efficiency using the anionic surfactant for air-water flow. His work leaves the scope for future work for determination of foam stability in presence of oil, which is aimed in this current study.

Kawale (2012) studied the influence of dynamic surface tension on foams: application in gas well deliquification. His work includes the influence of dynamic surface tension (DST) and surface tension (ST) on foamability in a customised setup. He used anionic (SDS), cationic (CTAB), non-ionic (Brij 30) and a commercial surfactant (Trifoam Block 820). He also tested his results in presence of Salt (NaCl). He worked on finding the correlation between the dynamic surface tension and the foamability of surfactants. Dynamic Surface tension was compared through Rosen parameter which was linked to foamability and foam density of surfactants. Later, Van Nimwegen (2015) studied the effect of surfactants on gas-liquid pipe flows in his dissertation. He analyzed the effect of surfactant on liquid unloading by injecting different types of surfactants and observing the flow behavior and pressure drop. Several parameters were evaluated like gas and liquid flow rates, surfactant concentration, surfactant type, pipe inclination and diameter. As a result of foam generation, he observed reduced liquid film density causing the film to move upward and lowering pressure gradient and fluctuations.

A few of other studies can be considered as good resources for foam lift. Barreto (2016) studied the effects of foamer delivery location on horizontal wells deliquification. She used an anionic surfactant, at a fixed concentration above CMC with continuous surfactant injection. She obtained improved unloading results in the lateral section when an external source of agitation was added. However, the foamer showed significant liquid unloading improvement in the vertical section for all cases. She also noticed that presence of surfactant can reduce or fully eliminate the severe slugging and improve the unloading. Later, Gcali (2017) studied the efficiency of surfactant batch treatment in horizontal gas wells. Her work includes the efficiency of the use of surfactants in batch configuration without any form of downhole agitation in horizontal gas wells. Like Gcali's method, the current study is focused on batch operation.

The above mentioned studies were carried out with air-water flow systems. But as the fraction of condensates in liquid composition increases, the foam lift efficiency reduces, which is evaluated in the current study. The study includes laboratory experiments, which are carried out with various surfactant types, surfactant concentrations, and water/condensate volumetric fractions. Initially, all the types of surfactants are characterized by static and dynamic surface tension measurements. Foam lift assessment is then completed using a liquid unloading column that serves as a replica for a gas well. Foam lift efficiency is determined by measuring the foam density, foam half-life, and unloaded liquid. All the experiments are conducted at atmospheric conditions. The outcome of this work sheds lights on the performance of a given surfactant type in the presence of varying amounts of condensates. The findings are intended to contribute as an operational guideline for the selection of surfactants to unload gas wells with varying water cuts.

Experimental Procedure and Setup

The performance of the surfactants depends on two important properties of surfactant solutions, foamability and foam stability. Foamability of surfactants largely depends on their surface tension and critical micelle concentration (CMC). Foam stability is a function of the liquid phase composition, the molecular structure and packing of surfactant, and other properties like temperature, pressure of the involved system.

At the first stage of this study, the static and dynamic surface tension of different types of surfactants were tested. The surfactant solution samples were prepared using deionized water and varying concentrations of surfactant. Static surface tension measurements were performed by Dynamic Contact Angle (DCA) analyzer (Wilhelmy plate). Using these measurements, CMC values were determined for a particular surfactant. Dynamic surface tension measurements were obtained using bubble pressure tensiometer.

These measurements provide a way to compare the performances of different surfactants in water. Dynamic surface tension is nothing but the time dependent value of static surface tension. The static surface tension value at a particular concertation is the value observed after equilibrium is established. The measurement of dynamic surface tension can be used to better understand the mobility of surfactant molecules in the solution and to compare the rate of reduction of surface tension for different surfactants, with different concentrations. The critical micelle concentration (CMC) is an important characteristic for surfactants. As the concentration of surfactant in the solvent (water) increases, the surface tension of the solvent decreases. Once the liquid surface is fully overlaid with surfactant molecules, aggregates i.e. micelles start forming in the bulk of liquid. The transition concentration at which micelles start to form is described as the CMC. Surface tension of the solution is not noticeably reduced further above CMC. Gcali studied the surfactant performance and concentration effects. She observed increases in foam production and liquid unloading by increasing the concentration below CMC. Beyond CMC however, she did not observe much improvement in terms of liquid unloading. Therefore, for this study three concentrations were selected for the unloading tests, below CMC, at CMC and above CMC. The surfactant concentrations are considered in parts per million (ppm).

Deionized water was used as the aqueous phase. The light hydrocarbon used for this study was Isopar L model oil, supplied by ExxonMobil Chemical Company. The oil has a specific gravity of 0.76, molecular weight of 162 g/gmol, pour point of -69 °C, flash point of 62 C, and the boiling point is 190°C. The properties of this oil simulate the characteristics of a gas condensate, making the comparison more realistic.

Surfactant Types

Surfactants are classified as anionic, cationic, amphoteric (or zwitterionic) and nonionic by presence of charged group in their head. When dissolved in water, they tend to concentrate at the interface. In an aqueous solution, the hydrophilic group of anionic surfactants dissociates into anions, hydrophilic group of cationic surfactants dissociate into cations, and hydrophilic group of amphoteric surfactants dissociates into anions and cations, depending on the pH. Types of surfactants included for this study are anionic, cationic, nonionic, amphoteric and two commercial surfactants provided by Halliburton. A brief description of the surfactants is as follows:

Anionic:
The anionic surfactant selected is 'Alpha Olefin Sulfonates' i.e. AOS. The sample is supplied by Stepan Company under the name POLYSTEP A-18. Depending upon the number of carbon atoms, the molecular weight of AOS can be in the range of 298.42 - 344.49. (average of 326 g/gmol). The concentration of surfactant is 39 %. The Molecular Formula for the surfactant is C_nH_{2n-1} SO_3Na (n = 14 - 16).

Commercial Surfactant 1
The first commercial surfactant provided by Halliburton (Multi-Chem Group LLC) is coded as MC 6400029 (Witcolate 1276). It is comprised of anionic and non-ionic type of surfactant mixture with isopropanol as a cosolvent. The concentration of surfactant is in the range of 40-80%.

Commercial Surfactant 2
The second commercial surfactant provided by Halliburton (Multi-Chem Group LLC) is coded as MC MX 4-1580. It is of anionic and non-ionic type of surfactant mixture with isopropanol as a cosolvent. The concentration of surfactant is in the range of 60-90%.

Cationic:
The cationic surfactant selected is Cetyl Trimethyl Ammonium Bromide (CTAB). The molecular weight of CTAB is 364.45 g/mol. Linear Formula is $CH_3(CH_2)15N(Br)(CH_3)_3$ i.e. $C_{19}H_{42}NBr$. The concentration of surfactant is 100%, and it was used in 40% weight concentration for this study.

Amphoteric:
The amphoteric surfactant selected is AMPHOSOL LB. The sample is provided by Stepan Company. The concentration of surfactant is in the range of 25-40 %.

Non-ionic:
The non-ionic surfactant selected is MAKON DA-9. The sample is provided by Stepan Company. The chemical description is Decyl Alcohol Ethoxylate. The surfactant is used at 100% concentration.

Figure 1 shows the static surface tension measurements for all of the above surfactants at varying concentrations. These measurements have also been used to determine the surfactant CMC values. The CMC is estimated as the concentration where the slope of concentration vs. surface tension changes. These CMC values are presented using the dashed lines with varying colors. The measurements were obtained at room temperature (25°C) with deionized water solution. Note that CTAB (C) means CTAB with cosolvent isopropanol, and CTAB (NC) means CTAB without cosolvent. Also, surface tension tests using DCA (i.e. Wilhelmy plate) are not available for water-oil mixtures. Table 1 shows the summary of different types of surfactants tested, and their respective CMC values.

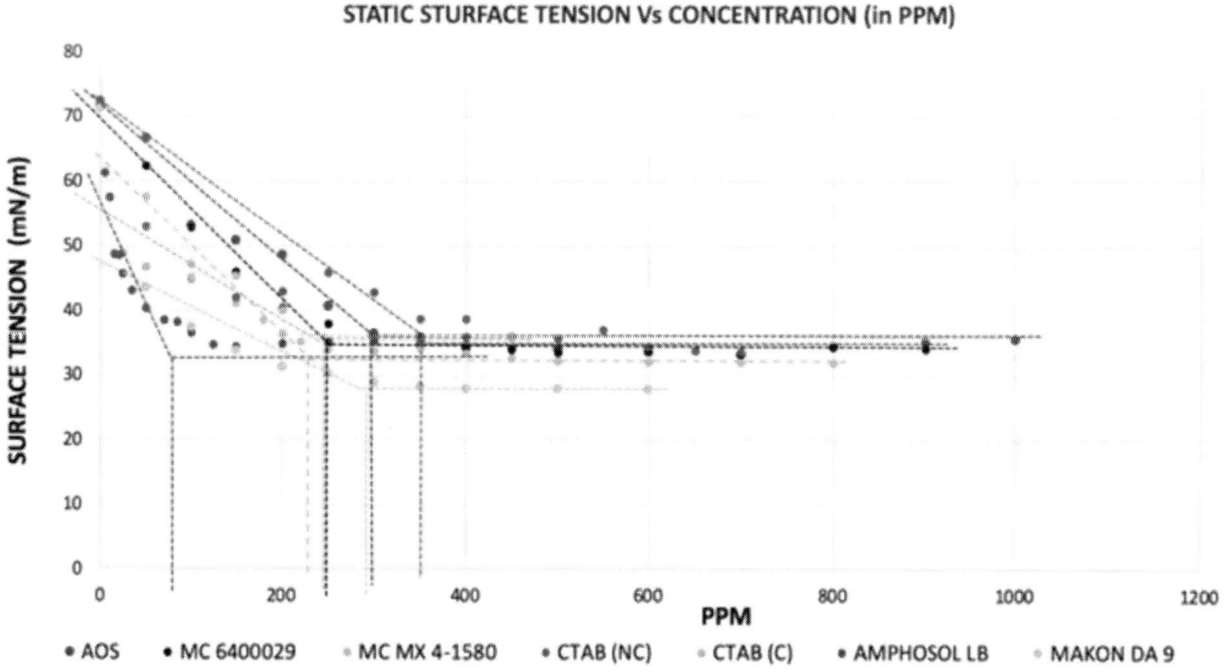

Figure 1. Static Surface Tension Measurements for Various Surfactant Types

Table 1: CMC values of surfactants

Surfactant Name	Type	CMC in PPM
POLYSTEP A-18 (Alpha Olefin Sulfonates)	Anionic	300
Commercial Surfactant 1: MC 6400029 (Witcolate 1276)	Anionic + Nonionic	250
Commercial Surfactant 2: MC MX 4-1580	Anionic + Nonionic	220
CTAB (Cetyl Trimethyl Ammonium Bromide)	Cationic	350 without cosolvent (NC) 250 with cosolvent (C)
AMPHOSOL LB	Amphoteric	70
MAKON DA-9	Nonionic	300

Figure 2 shows the dynamic surface tension measurements obtained at concentrations close to CMC for different surfactant types. These measurements show the transient behavior of the surfactant molecules at the liquid-gas interface. The case of water with no surfactant is also included in the plot. As observed, different surfactants have vastly different surface ages. The amphoteric surfactant takes the longest to perform, while the nonionic one acts the fastest to reduce the surface forces.

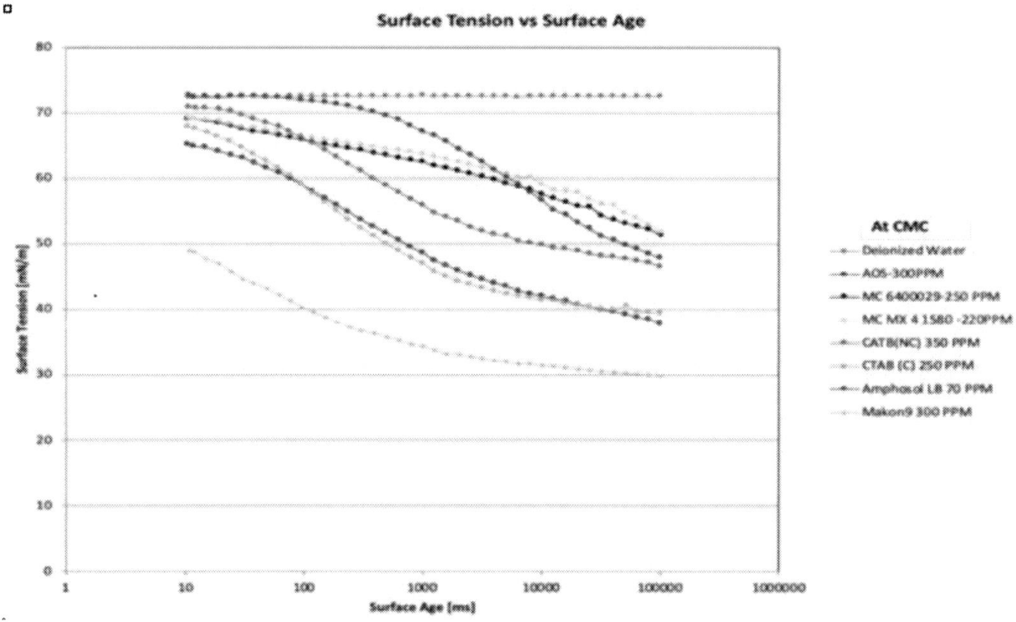

Figure 2. Dynamic Surface Tension Measurements for Various Surfactant Types

Unloading Column Setup

The experimental setup consists of a vertical production column, connecting arm, and a collection column. The production column is of dimensions of 7 cm (2.75") outer diameter (OD), 93.5 cm (37") height, and 1 cm thickness. The collection column is of dimensions of 17.7 cm (7") OD, 61.5 cm (24.2") height, and 0.5 cm thickness. The Connecting arm is 54 cm (21") long, 7 cm (2.75") OD and 1 cm in thickness. Material of the setup is Acrylic Plastic. High-pressure N_2 cylinders (2000 psi) are used as a source of pressurized gas, connected to the bottom of the production column. The connecting tubing for N_2 supply to the foam production column is of ¼ inch Silicon tubing. Foam is generated by sparging N_2 into the prepared surfactant solution in the production column. An inline pressure regulator is used to limit the downstream pressure to 200 psi maximum (so that it will not exceed the maximum allowable pressure of downstream pipe). The flow regulator is installed after the pressure regulator to control the flow rate. A flow meter is installed after the flow control valve to maintain the flow rate of 10 lit/min. This translate to an approximate superficial gas velocity of 0.3 ft/s for all of the conducted tests. All the deliquification experiments are carried out at atmospheric pressure and room temperature of 25°C.

Equipment were connected properly, making sure no leakage of N_2 and/or liquid is taking place, particularly at the joints like N_2 connection to the production column and production column to connecting arm. In a regular experiment, the 300-ml liquid mixture, with surfactant of a known concentration was poured into the production column. In the next step, N_2 supply of 10 lit/min was allowed to enter the column. As the N_2 was sparged into the production column, the foam generation starts. Foam started flowing to the top of the column and then through the connecting arm and finally collected at the collection column.

In most of the experiments, the foam unloading from the production column took place within the first few minutes. During this time, the initial foam production was seen continuous, then in the form of slugs or an annular flow pattern, where N_2 passes through the central portion and a foam layer flows upwards along the sides of the production column. The N_2 inlet valve was closed to stop the flow, when the foam flow was seen to stop reaching the top of the production column and getting in the collection column.

A stopwatch was used to record the time from the moment that foam production is started. The foam height, liquid drainage and the final liquid level readings in the collection column were noted down every minute for 60 minutes. For the cases with no oil, sometimes foam decay persisted beyond 60 minutes. For the cases with oil, the observed foam decay times were much shorter compared to the cases without oil. Finally, the remaining surfactant solution in the production column and the unloaded liquid (after foam decayed) in the collection column were measured using the measuring cylinders.

Figure 3 shows the schematic of the liquid unloading column setup. The foamability of different surfactants were tested using this setup. The first tests were for a base line case of liquid unloading without any surfactant. Tests were then repeated for each surfactant type at three concentrations i.e. below CMC, at CMC, above CMC. The results were compared for four water cuts, including 100% water, 90% water, 75% water, 50% water as the liquid phase.

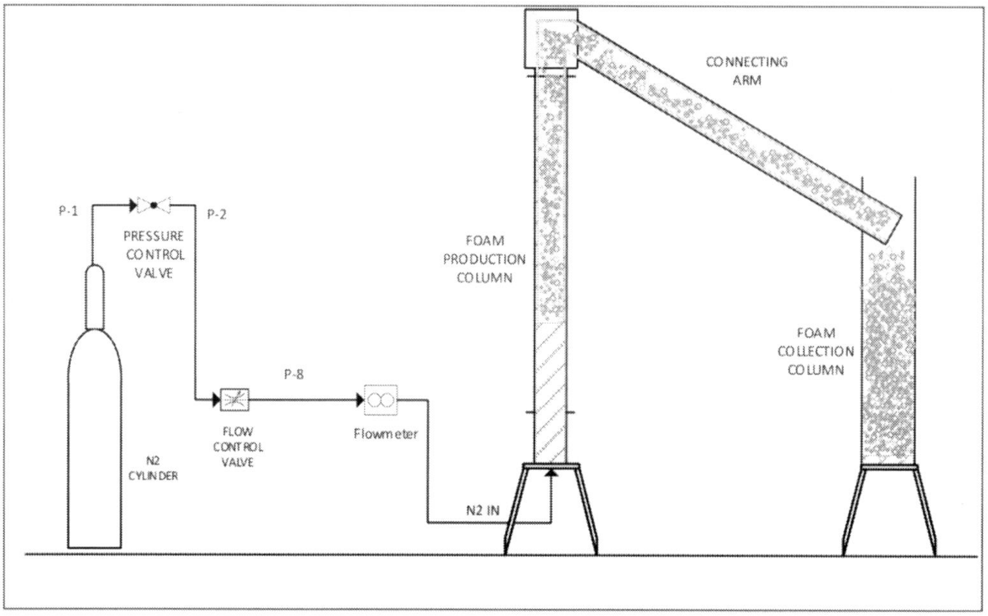

Figure 3. Schematics of Liquid Unloading setup

Results and Data analysis

The tests were conducted for various surfactant types, three surfactant concentrations, and four water cuts. At each test, the initial liquid volume at the production column was kept at 300 ml. During the foam decay time, the changes in foam and liquid height in foam collection column with time were recorded. After the conclusion of foam breakage, the final volume of unloaded liquid was also recorded to evaluate the surfactant unloading efficiency. The experimental results are presented in this section.

Figures 4 and 5 show the changes in foam height and liquid height with time, for three tests conducted with AOS as the surfactant and no oil in the liquid phase. The tests are completed with surfactant concentrations of 100, 300, and 500 ppm. Since the production time takes shorter than 3 minutes, the foam height peak at around 3 minutes for all cases. Then, it gradually decreases with foam breakage until the foam is completely collapsed. Foam height is recorded at every minute for 60 minutes duration. On the other hand, the liquid height gradually increases with time as the liquid drains out of the foam structure. The cases with concentrations of 300 ppm (around CMC) and 500 ppm (above CMC) show significantly larger heights of generated foam compared to the case of 100 ppm (below CMC). As a result, these two cases also result in larger liquid heights and better unloading efficiencies.

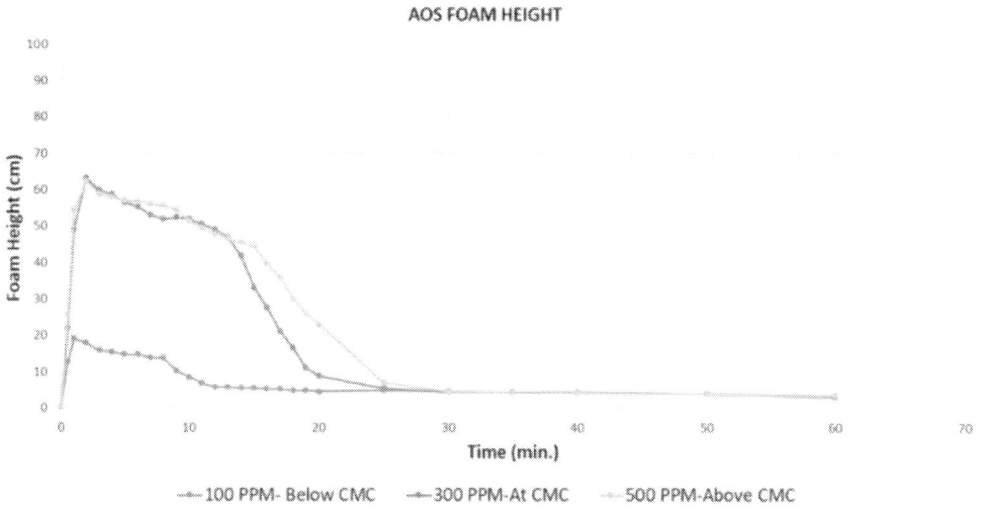

Figure 4. Changes in Foam Height with Time with AOS as the Surfactant and No Oil

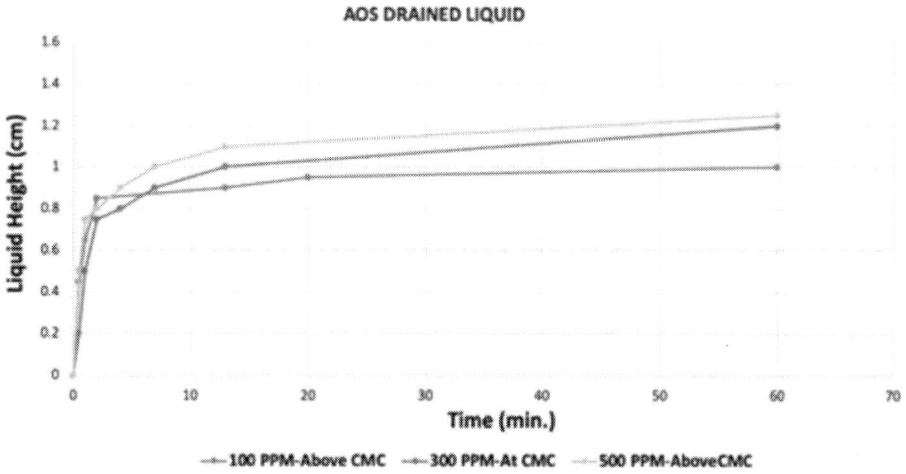

Figure 5. Changes in Liquid Height with Time with AOS as the Surfactant and No Oil

Figures 6 and 7 show the foam height and liquid height trends with time for the three conducted tests with AOS, this time with 25 % oil fraction in the liquid phase (75% water cut). Similarly, the results are presented with surfactant concentrations of 100, 300, and 500 ppm. These cases were selected to show an example of the effects of oil presence on foam generation. These graphs show substantial decreases in foam generation as compared to the case with no oil, shown in Figure 4. Also, the generated foam is seen to collapse very quickly. The liquid unloaded is also reduced significantly due to presence of oil.

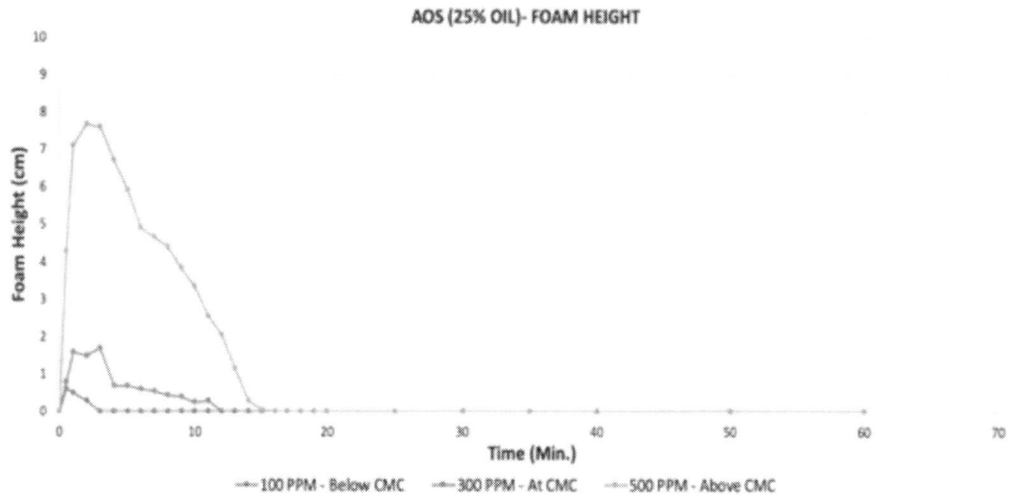

Figure 6. Changes in Foam Height with Time with AOS as the Surfactant and 25% Oil

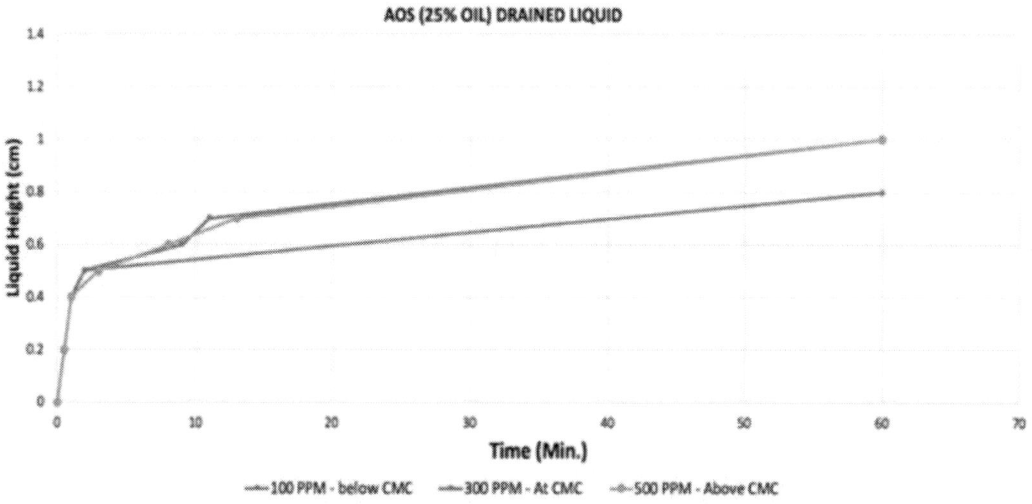

Figure 7. Changes in Liquid Height with Time with AOS as the Surfactant and 25% Oil

The results obtained in terms of unloaded liquid in the collection column and remaining liquid in the production column are a good indicator of foam lift efficiency for the conducted tests. As mentioned, the initial liquid volumes in the production column were fixed at 300 ml. Figure 8 shows the volumes of unloaded liquid, accumulated at the collection column for all the tests conducted with no oil in the liquid phase. The results are shown for all the surfactant types at concentrations below CMC, around CMC, and above CMC. As presented, no liquid unloading is observed in the absence of surfactant, for all the cases with varying water cuts. All surfactant types show acceptable liquid unloading performances in absence of oil, except for the amphoteric one (Amphosol LB). The efficiencies for other surfactants are very high and increase even more for concentrations around or above CMC, exceeding 80% of the initial liquid volume for all cases. The best performance belongs to Makon above CMC with around 95% unloading.

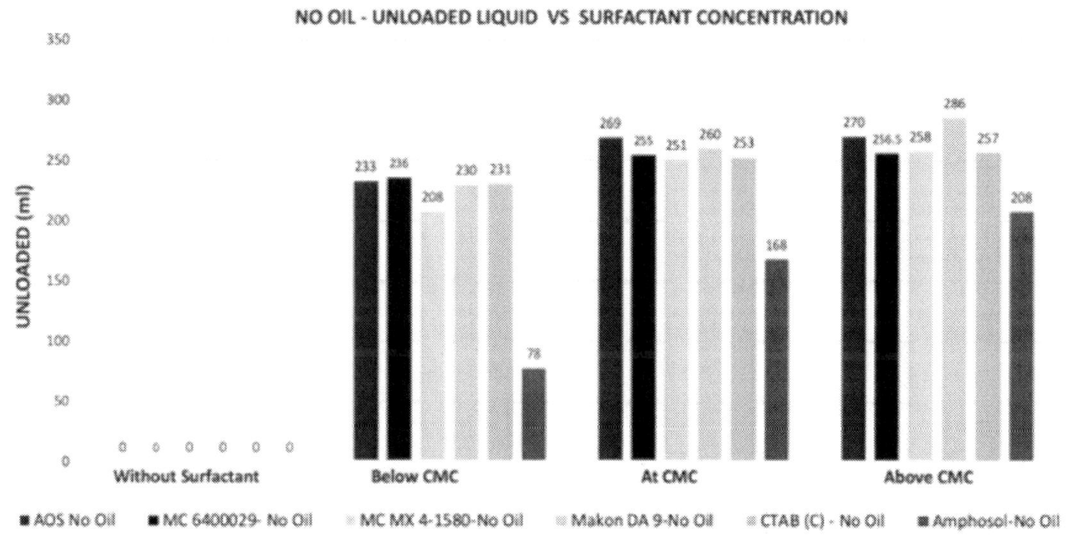

Figure 8. Unloaded Liquid Volumes for Different Surfactants with No Oil

Figures 9 and 10 show the unloaded liquid volumes for all surfactant types and concentrations with 10% and 25% oil fraction, respectively. Again, no unloading is observed for the case with no surfactant, and the results are reported for concentrations below CMC, around CMC, and above CMC for each surfactant. Unloading performances of all surfactant types are seen to reduce, compared to the case with no oil. The reason is the reduced foam generation in presence of oil, considering that the foam structure is the primary means for liquid unloading. Makon DA 9, AOS, and CTAB are seen to maintain relatively good performances for 10% and 25% oil fraction cases, with smallest drops in unloaded liquid volumes. However, the unloading efficiencies exceed 50% for almost all cases (around or above CMC) despite the increased oil fractions, showing that foam lift is still beneficial in well deliquification.

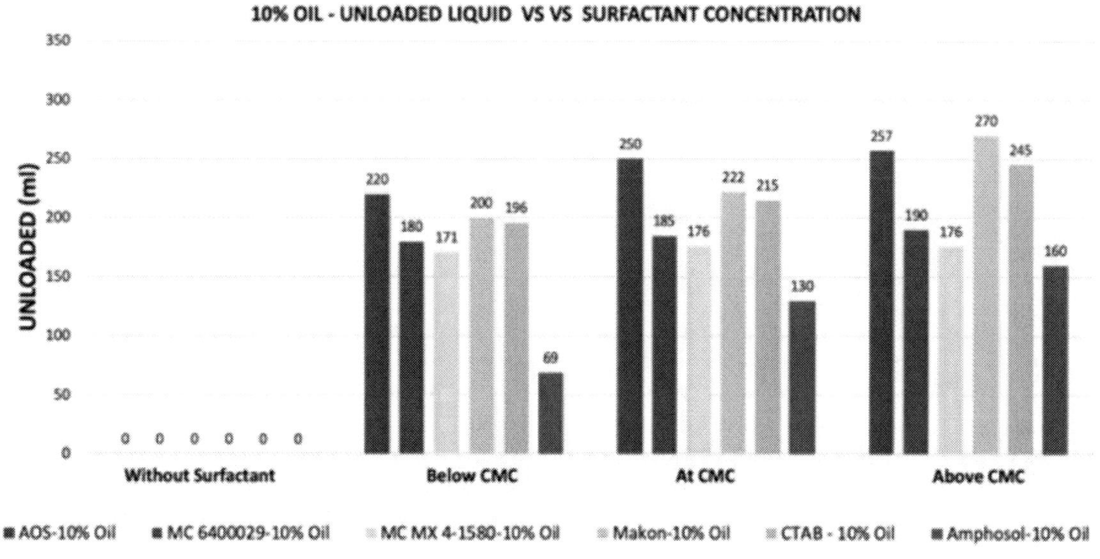

Figure 9. Unloaded Liquid Volumes for Different Surfactants with 10% Oil Fraction

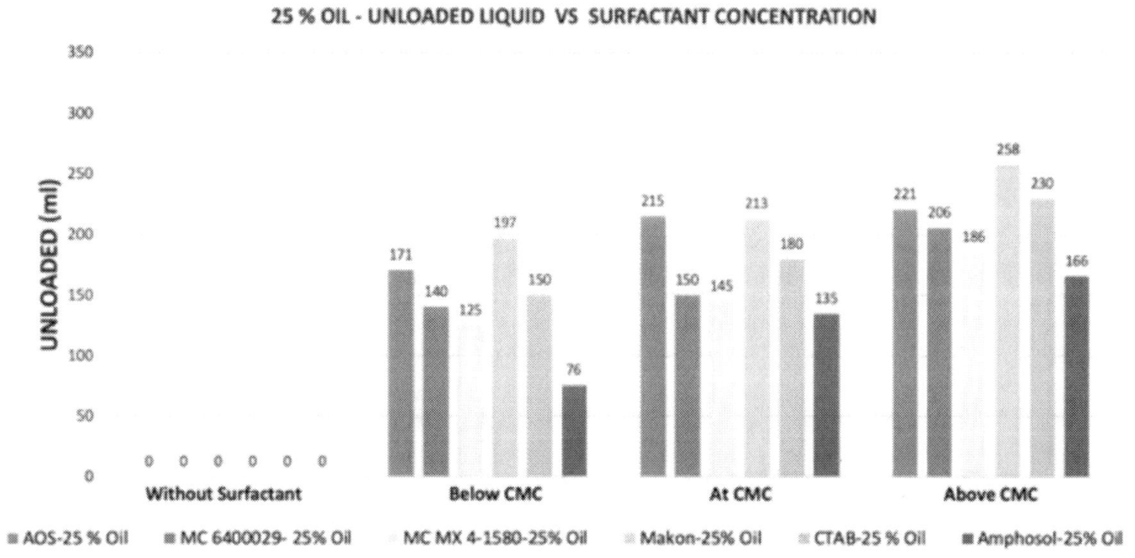

Figure 10. Unloaded Liquid Volumes for Different Surfactants with 25% Oil Fraction

Figure 11 shows the unloaded liquid volumes for all surfactant types and concentrations with 50% oil fraction in the liquid phase. The unloaded liquid volumes are significantly decreased for this case, showing the dependence of foam lift on fluid water cut. In some cases, particularly with Amphosol, surfactant becomes completely inefficient. Also, the concentration effects are seemingly maximized, favoring the cases with above CMC concentrations. The commercial surfactants show relative improvement for the above CMC case. Also, Makon DA 9 is still maintaining an acceptable performance, followed by CTAB. Further study of the surfactant chemistry is required to better understand the behavior of different surfactant types.

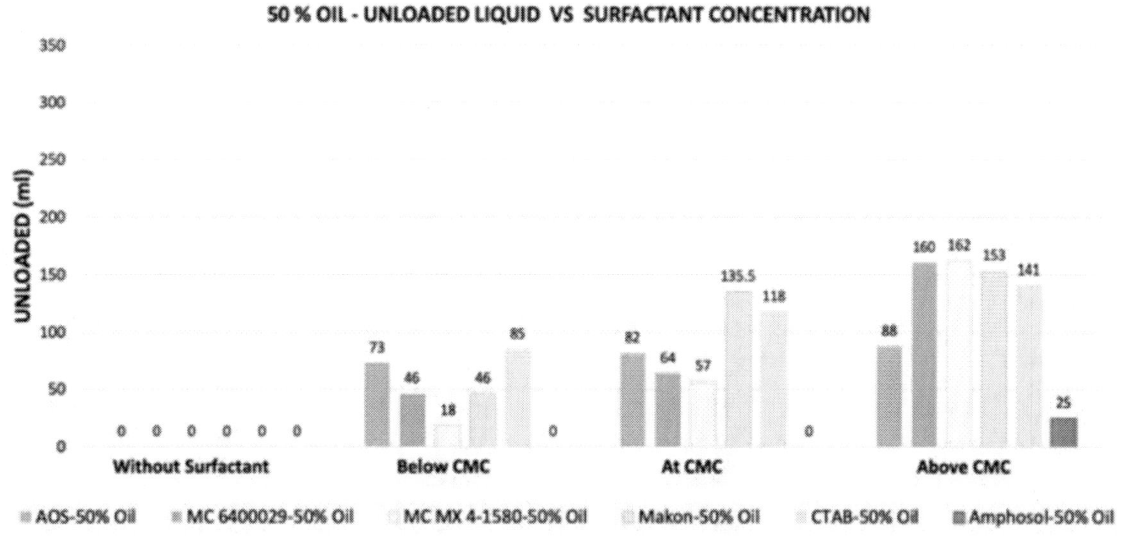

Figure 11. Unloaded Liquid Volumes for Different Surfactants with 50% Oil Fraction

Conclusions

It was clearly observed from the results that presence of oil has reduced the foamability of all the surfactants. The foam volume or maximum foam height was affected substantially in presence of oil. Out of all the tested surfactants, Makon DA 9 (non-ionic) kept a good performance for all the cases. AOS (anionic) showed good results for 100% water case, but its performance dropped in presence of higher oil fractions. CTAB (cationic) showed good performance for 50% oil case. Both the Commercial surfactants showed good results at higher concentrations (above CMC) for 50% oil case.

The future scope of this study can include conducting tests at high temperature and high pressure environment to better represent a well downhole. The tests can also include varying gas flow rates to predict its effect on surfactant unloading efficiencies. The outcomes of this work help us better understand the performance of a given surfactant type in presence of condensate. The findings can contribute as an operational guideline for the selection of surfactants to unload gas wells with varying water cuts.

References

"Influence of Dynamic Surface Tension on Foams: Application in Gas Well Deliquification" - A dissertation by Durgesh Kawale (MSc Delft University of Technology 2012).

"Experimental Study and Modeling of Effect of Surfactants on Liquid Loading in Vertical Pipes" - A dissertation by Abdulkamil Ajani (Ph.D. at The University of Tulsa 2014).

"The Effect of Surfactants on Gas-Liquid Pipe Flows" – A dissertation by A.T. van Nimwegen (Ph. D. at Delft University of Technology 2015).

"Effect of Foamer Delivery Location on Horizontal Wells Deliquification" - A dissertation by Carolina Barreto (M.S. in P. E. at The University of Tulsa 2016).

"Efficiency of Surfactant Batch Treatment in Horizontal Gas Wells" - A dissertation by Cikida Gcali (M.S. in P. E. at The University of Tulsa 2017).

577554 A CFD Based High-Dimensional Deep Neural Network Surrogate Model for Oil/Water Separation in Horizontal Oil Production Pipelines

A CFD based high-dimensional deep neural network surrogate model for oil/water separation in horizontal oil production pipelines

-Kuochen Tsai and Ligang Lu, Shell International Exploration & Production, Inc.

Deep neural network (DNN) model has been proven to be highly successful in modeling computational fluid dynamics data for oil/water separations in oil production pipelines in 4-dimensional parametric space [Ref 1]. Further development demonstrates the feasibility for a 9-dimensional DNN model with physical parameters that covers the entire range of operating conditions, including mixture velocity, water cut, initial droplet size, pipe diameter, fluid viscosity, density and surface tension.

The computational fluid dynamics (CFD) model is based on the Eulerian Interfacial Area Exchange (IAC) model of Ishii and Kim [Ref 2]. The CFD model development was presented in 2017 with comprehensive validation with published experimental studies [Ref 3, 4, 5]. The validations showed the accuracy and scale-up capability of the CFD model.

The combination of DNN and CFD is a powerful tool that can be further integrated with a web model using public domain software (Python and Dash Plotly). The result is a general- purpose, user friendly real-time web model for oil/water separation predictions.

Similar ideas can be extended for many other types of physics-based simulation tools including finite element models and process simulators. The concept of extracting 1-D or 2-D subset data from a 3-D model is highly flexible and can be a powerful tool for converting time-consuming simulations into simple real-time general-purpose web models.

References

[Ref 1] K. Tsai and L. Lu, "A Deep Learning Neural Net Model for Oil/Water Separation in Oil Production Pipelines", Paper 152g, 2019 AIChE Annual Meeting, Pittsburth, PA.

[Ref 2] Wu Q., Kim, S. and Ishii, M.: "One-group interfacial area transport in vertical bubbly flow," *International Journal of Heat and Mass Transfer* (1998), 41(8–9), 1,103–1,112

[Ref 3] K. Tsai, "Modelling the separation of oil and water in pipelines", Paper 452a, 2017 AIChE Annual Meeting, Minneapolis, USA

[Ref 4] Chesters, A. K.: "The modelling of coalescence process in fluid–liquid dispersions: A review of current understanding," *Chemical Engineering Research and Design* (1991), 69 (A4), 259–270

[Ref 5] Simmons, M. J. H. and B. J. Azzopardi: "Drop size distribution in dispersed liquid–liquid pipe flow,"*International Journal of Multiphase Flow* (2001) 27(5), 843–859

[Ref 6] Fairuzov Y. V., Arenas-Medina, P., Verdejo-Fierro, J. and Gonzales-Islas, R.: "Flow pattern transitions in horizontal pipelines carrying oil-water mixtures: Full-scale experiments," *Journal of Energy Resources Technology* (2000) 122(4), 169–176

576181 Polymer Drag Reduction in Taylor-Couette Flow: Routes for Improved Shear Stability

Adding high molecular weight polymers to fluids can significantly reduce frictional drag when in turbulent flow. However, the good drag reduction performance is often limited by the time- and shear-dependent stability of the polymer, with polymer degradation leading to reduced drag reduction. Reduced performance can be partially mitigated by frequent dosing of the transport fluid, although routes to resist degradation are desired. Currently, the mechanisms of polymer degradation are poorly understood, and therefore routes to overcome such issues have resulted in little clarity.

In the current study, the drag reduction performance of a high molecular weight (> 1 MDa) polymer, Poly(2-dimethylamino)ethyl methacrylate) methyl chloride quaternary salt (Poly DMAEMAq), was measured using a Taylor-Couette flow geometry (double-walled gap, TA Instruments DHR-II) up to *Reynolds number* 5,500. From a stress ramp test, the onset of drag reduction (relative to the solvent) occurred at a critical shear rate of ~840 s^{-1}. The polymer concentration to attain the highest drag reduction of 19% was 40 ppm, as shown in Fig. 1A. Further increases in polymer concentration led to no further enhancement. While demonstrating good drag reduction (maximum limited by *Re,*), the highly charged (cationic) polymer exhibited poor shear stability (Fig. 1B), with the effect of drag reduction diminishing within the first 100 s of the test at a range of Reynolds numbers (1,500 – 5,000). Varying the polymer concentration from 1 ppm to 40 ppm had negligible influence on the shear stability, and increasing the shear rate up to ~5,300 s^{-1} (*Re* = ~5,000) resulted in a rapid drop off in performance.

To overcome the poor shear stability, the current study has considered the influence of polymer-surfactant (CTAB) co-addition and solution salinity. The surfactant CMC in the presence of an equimolar concentration of sodium salicylate (NaSal) was measured to be ~50 ppm. Compared to a polymer solution only, the co-addition led to an initial slight increase in the drag reduction at shear rates < 1,500 s^{-1}, which can be attributed to the drag reduction by surfactant worm-like micelles. At higher shear rates, the surfactant no longer provides a contribution to the overall drag reduction with values comparable to the polymer only. However, when measuring the shear stability of the drag reducing agent, slight improvements in the time-dependent performance were detected. This effect is thought to relate to the increased salinity of the solution and a change in polymer conformation (i.e. coiling through charge screening). The research discusses the potential synergistic effects between polymer and surfactants/solution salinity to optimize the potential of the polymer for drag reduction applications.

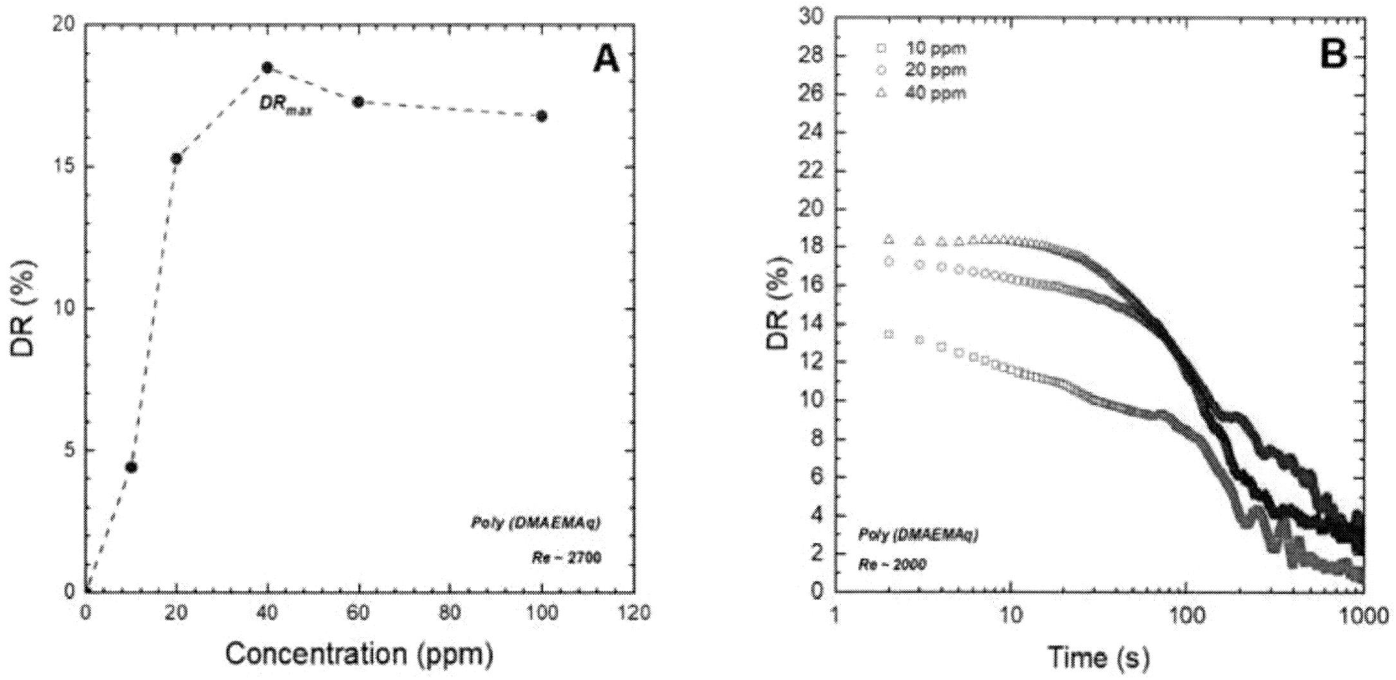

563567 An Automated Microfluidic System for High Throughput Mechanistic Study of Asphaltenes Deposition and Dissolution in Porous Media Using *in Situ* Spectroscopy

Asphaltenes, known as 'cholesterol in petroleum', attributed to the viscous properties of the crude oil. The precipitation and deposition of asphaltenes could lead to the fouling and impairment problem in petroleum industry. The motivation to tackle these problems triggered great interests in both academic and industrial research. However, the study of asphaltenes could be laborious and time-consuming due to the broad definition of asphaltenes. Defined as materials that are soluble in aromatic solvents and insoluble in alkane solvents, asphaltenes molecules have a rather complex molecular structure: aromatic core with heteroatoms (e.g., N, O, and S) and metals (e.g., V, Ni, and so on).[1–4] These facts result in requirement for high throughput experimentation in asphaltenes study.

In the present work, we designed and assembled an automated microfluidic system to achieve the high throughput study of asphaltenes deposition and dissolution in porous media. The experimental period was decreased from days and weeks to minutes and hours when compared with our previous version of micro-packed bed reactor. The Peltier module enabled the swift temperature change of the microsystem at the rate of 1.0 K/s. The *in-situ* characterization window of the microsystem enabled the access to implement the characterization of the micro-packed bed region during the asphaltenes deposition and dissolution process by Raman spectroscopy for the first time. The acquisition of Raman mapping of the 4.5-μL micro-packed bed region took about 21 min to finish. As a result, one would obtain 5 maps during asphaltenes deposition and dissolution process. The result of bed occupancy provided the concentration of asphaltenes onto the porous media, while the sheet size information of asphaltenes provided the molecular level information about asphaltenes. The results will be presented and explanted to provide the insights toward the mechanisms of asphaltenes deposition and dissolution.

Reference:

(1) Mullins, O. C. The Asphaltenes. *Annu. Rev. Anal. Chem.* **2011**, *4* (1), 393–418.

(2) Gray, M. R.; Tykwinski, R. R.; Stryker, J. M.; Tan, X. Supramolecular Assembly Model for Aggregation of Petroleums Asphaltenes. *Energy Fuels* **2011**, *25* (7), 3125–3134.

(3) Hu, C.; Morris, J. E.; Hartman, R. L. Microfluidic Investigation of the Deposition of Asphaltenes in Porous Media. *Lab Chip* **2014**, *14* (12), 2014–2022.

(4) Pinho, B.; Minsariya, K.; Yen, A.; Joshi, N.; Hartman, R. L. Role of HZSM-5 Aluminosilicates on Asphaltenes Deposition by High-Throughput in Situ Characterizations of a Microreservoir. *Energy Fuels* **2017**, *31*, 11640–11650.

563922 Modeling Asphaltene Deposition in Production Pipelines

Asphaltenes are heavy fractions that are dissolved in crude oil. The sudden change of either temperature, pressure, or composition affects the phase stability of crude oil and leads to the precipitation of asphaltenes. After precipitation occurs, asphaltenes begin to deposit on different locations along the flow of crude oil, including pipelines, well tubing, near wellbore region, and inside the formations. As deposits build up, they begin to cause a significant reduction in production rates. Risks associated with these asphaltenes related flow assurance problems need to be accurately quantified in order to develop appropriate mitigation strategies. In this study, a simple and robust model based on sound physics is developed and tested for field-like conditions by scaling-up the diameter and adjusting the flow rate. The model proved its ability to accurately predict the deposition rate.

561892 Effect of Asphaltenes on the Structure and Surface Properties of Wax Crystals in Waxy Oils

As the temperature of waxy crude oils decreases, the dissolved wax molecules precipitate because of supersaturation. When the precipitated wax crystals reach a certain amount, cross-links occur among wax crystals, gradually forming a three-dimensional network, which leads to the gelation of crude oils. Asphaltenes, as the most polar component of crude oils, are able to interact with wax molecules during the wax crystallization, thus changing wax precipitation properties and wax crystal structure, and affecting the gelation behavior of crude oils. In the present work, the effect of asphaltenes on waxy model oil systems was carried out from four aspects: macroscopic wax precipitation and gelation properties, mesoscopic particle size and distribution of wax crystals, microscopic lattice structure, and zeta potential on wax crystal surfaces.

The model oils were prepared using mineral oils and mixed paraffines with variable asphaltene concentrations (0~0.20wt%). The results of the differential scanning calorimetry (DSC) and rheological measurements showed that asphaltenes could inhibit wax precipitation, delay gelation, and weaken gel structure. This impact would be magnified with the increase of asphaltenes. Through the X-ray diffraction (XRD) analysis, it was concluded that the crystallinity index (CI) decreases with the asphaltene increase as well, indicating asphaltenes hinder the wax precipitation, which confirms the DSC results. Analysis on the gel structure of waxy oils was performed via the stress sweep test and scaling theory. It was found that as more wax crystals precipitate out, the gel structure changes into the weak-link regime from the strong-link regime, and the critical amount of precipitated wax corresponding to the transition point will increase when there are more asphaltenes.

Quantitative analysis on the morphology of wax crystals was conducted by means of polarized optical microscopy (POM), Image J software, and Nano Measure 1.2 software. As the asphaltene concentration rises, the average size of wax crystals gradually reduces, and the size distribution tends to smaller particle size; furthermore, it is obvious that the wax crystals transform from distinct acicular shape to the elliptical or punctiform. The aspect ratio and average perimeter of wax crystals decreases while the boundary box fractal dimension increases, demonstrating that the morphology of wax crystals becomes more complex. Based on the XRD pattern, the grain size of wax crystals was calculated by Scherrer equation. The result showed that the grain size presents a drop trend with the increase of asphaltenes. Since the grain is considered as the basic unit of wax crystals, it is inferred that the reduction of grain size is one of the reasons leading to wax crystals getting smaller.

The lattice structure of wax crystals was studied by the XRD and MDI Jade 6.0 software. As the asphaltene is more concentrated, the lattice parameters a and b rarely varies whereas the lattice parameter c significantly increases. According to the end-gauche defect, the increase of lattice parameter c indicates the extent of conformational disorder of wax crystals is strengthened, resulting in the weakening of gel strength, which shows up as the decline of yield stress and the delay of gelation.

Finally, the zeta potential of wax crystals was detected. Based on the Derjaguin-Landau-Verwey-Overbeek (DLVO) theory, it was observed that with higher asphaltene concentrations, the zeta potential of wax crystals enhances. More negative charges are absorbed on the surface of wax crystals, and larger electrostatic repulsion generates between wax crystals due to the electrostatic double layers. Therefore, the growth and aggregation of wax crystals are interfered, which counts against gelation.

This paper is supported by the National Natural Science Foundation of China (No.51534007 and No.51134006)

555733 Paraffin Deposition Modeling – Is the Current Methodology Sufficient?

Paraffin deposition in subsea pipelines is one of the greatest flow assurance challenges in offshore production, and reliable paraffin deposition models serve as vital guiding tools for field operations. Nearly all previous work simulate paraffin deposition using a "diffusion-limited" approach. That is, molecular diffusion is regarded as the primary deposition mechanisms and treated as the rate-limiting step in modeling practices. In this work, we present experimental observations that questions this approach. The results are a collection of findings from several systematic experimental investigations conducted by state-of-the-art flow loop facilities and a waxy condensate with 9.5% wax content. These data, along with recent findings from other publications, collectively indicate that the prevailing approach of modeling paraffin deposition is insufficient. We suggest further modeling efforts take other deposition mechanisms and the non-Newtonian feature of paraffin deposits into consideration.

567874 The Influence of Nanocomposite Pour Point Depressant on Wax Deposition of Waxy Oil

The Influence of Nanocomposite Pour Point Depressant on Wax Deposition of Waxy Oil

ABSTRACT:

With the advancement of deep-water oil and gas energy strategy, the flow assurance problem of various solid depositions in pipelines is becoming more severe as the heavy hydrocarbons produced from the offshore fields accompanied with sharply variable temperature of outflow environment. The nanocomposite pour point depressant (NPPD), which is developed by blending of nanocompostie and polymers, has been introduced to improve the fluidity of waxy oil and aroused tremendous attention. The modification effect of NPPD on the rheological of waxy oil and its interaction mechanism are widely investigated, however, the influence of NPPD on wax deposition process is not further discussed.

In this work, a new NPPD was synthesized based on the melt blending of organically modified nano-montmorillonite and ethylene/vinyl acetate copolymers. The high temperature gas chromatography (HTGC), cold finger and flow loop were chosen to investigate the variation of carbon number distribution and thickness of the wax deposition layer before and after NPPD modification. The crystal morphology and rheological behavior of undoped/doped waxy oil were characterized by polarizing microscope (POM) and rheometer. Based on the macro and micro results, the interaction mechanisms between NPPD and wax molecules were discussed systematically, which provided a better understanding of the effect of NPPD on wax deposition of waxy oil.

As is shown in Figure 1(A), for undoped system, with the increase of deposition time, the heavy component wax content (carbon number>=C31) of sediment progressively increases and the peak of chromatogram moves to higher carbon number. After addition of NPPD (50 mg/kg), this trend is enhanced obviously (Figure 1(B)), where the content of heavy component wax is almost twice larger than undoped system. Comparing the absolute mass of individual carbon number components, it is found that the deposits of doped waxy oil tend to have more high carbon number components (>=C33) than undoped waxy oil deposits (Figure 1(C)). The results of macroscopic rheology further inferred that the decrease of the bulk viscosity after NPPD modification facilitates the diffusion and deposition of wax molecules (Fick law), which is benefit to the deposition of higher wax fraction (>=C33). Furthermore, the loop experiments of undoped/doped waxy oil under laminar flow and turbulent flow indicates that the wax deposition thickness is decreased after addition of NPPD. However, it is noticed that the deposit thickness-time curves are almost overlapping at range of 100 to 200 mg/kg NPPD in the turbulent flow state (Figure 1D). The observation results from POM exhibit that wax crystal morphology of waxy oil is sensitive to the dosage of NPPD, which varies from 0.025mm (0 mg/kg) to 0.012mm (200 mg/kg). A plausible explanation for these results is that the introduction of NPPD which provides a nucleation template for wax crystal and modifies the morphology of crystal morphology, resulting in weaken of three-dimension network of wax crystals and decrease the deposition thickness layer. These conclusions provide a comprehensive theory analysis for the application of NPPD and the technology of flow assurance in waxy oil pipelines.

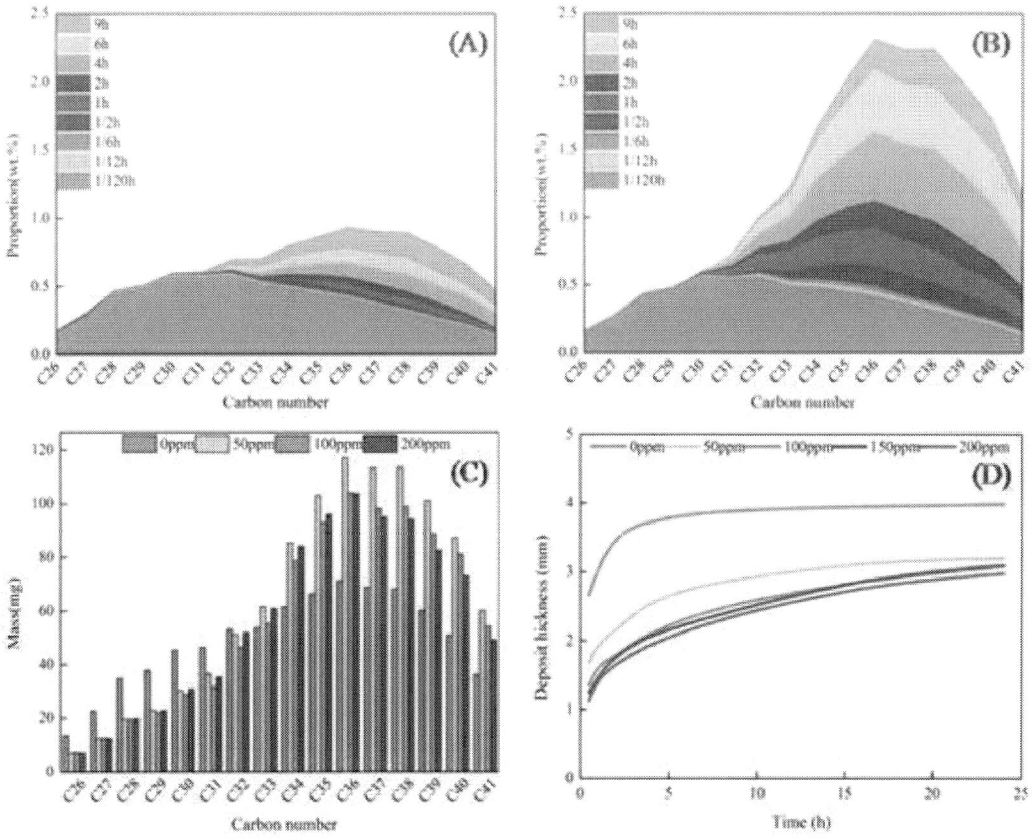

Figure 1. The carbon number proportion of undoped waxy oil in cold finger experiment (A); the carbon number proportion of doped waxy oil (50 mg/kg NPPD) in cold finger experiment (B); the absolute wax content of deposition layer in cold finger (C); the thickness of deposition layer in flow loop under turbulent (D).

Keywords: Flow assurance; Wax deposition; Waxy oil; NPPD

574596 Phase Equilibrium Conditions of Methane Hydrate in Presence Aqueous Solutions of $CaBr_2$, $MgBr_2$ and $ZnBr_2$

Phase equilibrium conditions of methane hydrate in presence aqueous solutions of $CaBr_2$, $MgBr_2$ and $ZnBr_2$

Dnyaneshwar R. Bhawangirkar[1,2], Jitendra S. Sangwai[2*]

[1]Department of Chemical Engineering, Indian Institute of Technology Bombay, Powai, Mumbai 400076, India

[2]Gas Hydrate and Flow Assurance Laboratory, Petroleum Engineering Program, Department of Ocean Engineering, Indian Institute of Technology Madras, Chennai 600036, India

Keywords: Gas hydrates; Inhibition; Flow Assurance; High pressure phase equilibria

ABSTRACT

Gas hydrates form when water molecules and suitably sized gas molecules, like methane, ethane etc. come in contact at low temperatures and high pressures. In oil and gas industry, the formation of gas hydrates usually occur in drilling operations, production wells and transportation pipelines; and can cause plugging, pipeline blockages and other operational problems. To address this issue, from a more practical point of view, inhibitors are used so as to prevent or delay the formation of hydrates.

Chemical inhibitors can also help in releasing methane gas molecules trapped in the form of hydrates deep in the ocean and in permafrost regions by shifting the pressure temperature conditions. It is necessary to establish a reliable experimental phase equilibrium (dissociation) data for gas hydrates in the presence of aqueous solutions of various inhibitors. Although experimental data has been reported in the open literature for methane hydrate phase equilibria in the presence of inhibitors like NaCl, KCl, $CaCl_2$, $MgCl_2$, $ZnCl_2$, NaBr, KBr; however information on some inhibitors is still limited. Therefore, the dissociation conditions of methane hydrate in presence of aqueous solutions of $CaBr_2$, $MgBr_2$, and $ZnBr_2$ are measured in this study. We have used four different weight percentages of $CaBr_2$, $MgBr_2$, and $ZnBr_2$ i.e., 1, 5, 10 and 15% such that their effect on the formation of methane hydrates can be studied. We have used an iso-chor pressure search method to generate the equilibrium data in the temperature and pressure ranges of (273.6-285.6) K, and (1.5-9.5) MPa, respectively. Results reveal that the satisfactory inhibition effect has been observed on the methane hydrate formation, causing the change in *P-T* conditions by shifting the equilibrium point to higher ranges. We also observed that the inhibition effect is more dominant at higher weight percentage of salts. This information will certainly help in oil and gas industry with new hydrate dissociation data on chemical inhibitors at various pressures, temperatures and at different concentrations. This signifies the effectiveness and usefulness of these salts used here in inhibiting methane hydrate formation for a smooth operation in transportation pipelines, production wells and in drilling operations.

559738 Equilibrium and Non-Equilibrium Phase Behavior of CO_2 and Crude Oil Interactions

The visualization and quantification of CO_2 and oil interactions give insight into the multiple mechanisms controlling CO_2 enhanced oil recovery processes. In this work, a high pressure high temperature full visual PVT system is used to observe and measure the equilibrium properties of CO_2 solubility in Bakken oil, swelling factor, and minimum miscibility pressure (MMP), together with the non-equilibrium properties of CO_2 condensation, light oil components extraction, and diffusion coefficient at elevated temperatures. The mechanisms at the different stages of CO_2 EOR process are identified and analyzed.

Firstly, excessive gaseous CO_2 is charged into the piston-equipped view cell coexisting with the pre-loaded Bakken oil. Three types of phase behavior can be captured under certain conditions including vapor-liquid (VL), vapor-liquid-liquid (VLL), and liquid-liquid (LL), which are common fluid types in the formation during CO_2 EOR process. In equilibrium process, the cell is pressurized stepwise. The volume of the swollen oil caused by the dissolution of CO_2 is recorded at each equilibrium pressure step, at which gas solubility and swelling factor can be calculated. Once the light components in Bakken oil start to be extracted into the liquid CO_2 phase, the volume of oil-rich phase decreases and a couple of extraction columns can be observed. The heavy components in oil are harder to be extracted so that the oil volume eventually reaches a plateau, where the pressure is recorded as the pseudo MMP and the system is at the near miscible condition. During the non-equilibrium process, the pressure increases continuously by moving the piston at the various rates until the pressure reaches the MMP. Finally, the pressure decay method is used to determine the diffusion coefficient between CO_2 and Bakken oil. It has been found that oil can be swollen by dissolving CO_2 at high pressures. The swelling factor increases with pressure during VL condition, where the EOR mechanism is mainly oil swelling effect. However, the swelling factor decreases with pressure during VLL and LL conditions, indicating a change in the main controlling mechanism to CO_2 extraction at the near miscible condition. As for the non-equilibrium process, the extraction pressure is found to be closely related to the moving rate of the piston. The condensing flow from CO_2 rich liquid phase to oil phase and the extracting flow from oil phase to CO_2 rich liquid phase have been filmed to demonstrate the EOR mechanisms. The effective diffusion coefficient in Period I, which is dominated by natural convection, is found to be three orders larger than that in Period II, which is mainly driven by molecular diffusion.

569760 Deciphering the Complex Fluid Phase Behavior in Nanoporous Materials

Deciphering the complex fluid phase behavior under confinement is critical in numerous energy and environmental related applications. For instance, the greater fundamental understanding of various reservoir fluids confined in geological nanoporous media could significantly advance the current technologies aimed towards enhancing the hydrocarbon recovery from unconventional resources (e.g., shales) and storing fluids such as carbon dioxide in these resources. In present investigation, a coupled grand canonical Monte Carlo (GCMC) and molecular dynamics (MD) simulation techniques were used to provide the fundamental explanation to a complex adsorption and diffusion behavior of various linear and non-linear organic moieties (including n-hexane, n-heptane, cyclohexane and methyl-cyclohexane) confined in silica nanopore(s). The adsorption and diffusion of linear and non-linear organic species exhibit stark differences largely influenced by the dominating intermolecular interactions originating from the internal pore surfaces, and the steric effects.

560189 Estimation of Three-Phase Relative Permeabilities for a Wag Process Using an Improved Ensemble Randomized Maximum Likelihood Algorithm

A modified ensemble randomized maximum likelihood (EnRML) algorithm has been developed to estimate three-phase relative permeabilities with consideration of hysteresis effect by reproducing the actual production data. Ensemble-based history matching uses an ensemble of realizations to construct Monte Carlo approximations of the mean and covariance of the model variables, which can acquire the gradient information from the correlation provided by the ensemble. A power-law model is firstly utilized to represent the three-phase relative permeabilities whose coefficients can be automatically adjusted until production history is matched. A damping factor is introduced as an adjustment to the step length since a reduced step length is commonly required if an inverse problem is sufficiently nonlinear. A recursive approach for determining the damping factor has been developed to reduce the number of iterations and the computational load of the EnRML algorithm. The restart of reservoir simulations for reducing the cost of reservoir simulations is of significant importance for the EnRML algorithm where iterations are inevitable. By comparing a direct-restart method and an indirect-restart method for numerical simulations, we optimize the restart method used for a specific problem. Subsequently, we validate the proposed methodology by using a synthetic water-alternating-gas (WAG) displacement experiment and then extend it to match laboratory experiments. The proposed technique has proved to efficiently determine the three-phase relative permeabilities for the WAG processes with consideration of the hysteresis effect, while history matching results are gradually improved as more production data are taken into account. The synthetic scenarios demonstrate that the recursive approach saves 33.7% of the computational expense compared to the trial-and-error method when the maximum iteration is 14. Also, the consistency between the production data and model variables has been well maintained during the updating processes by using the direct-restart method, whereas the indirect-restart method fails to minimize the uncertainties associated with the model variables representing three-phase relative permeabilities.

Cold Rotating Finger: Apparatus to Study Wax Deposition under Shear

G. Yaghy*⊥, T.V.J. Charpentier⊥, L. Fusi§, A. Neville† and D. Harbottle⊥

⊥ School of Chemical and Process Engineering, University of Leeds, Leeds, UK
† Institute of Functional Surfaces (iFS), School of Mechanical Engineering, University of Leeds, Leeds, UK
§ I2T3 – c/o: Dipartimento di Matematica "U.Dini", Universita´ di Firenze, Viale Morgagni 67/A, 50134 Firenze, Italy

***Corresponding Author**
E-mail: pm14gmy@leeds.ac.uk

ABSTRACT
A bench-top Cold Rotating Finger (CRF) is developed to study the effect of fluid shear on wax deposition. An inner rotating cylinder provides opportunity to study wax deposition on surfaces in the region of high fluid shear stress (and laminar/turbulent flows). A model system of linear n-alkane (C20 to C40) dissolved in dodecane to 9 wt% was used for the deposition tests. The bulk fluid temperature was set to 45°C and the temperature of the inner rotating cylinder to 5°C, with wax deposits showing a mass variation less than ~5% at constant rotational speed. The kinetic of wax deposition is non-linear, with an initial rapid deposition (approximately linear) followed by a slower rate as deposit mass tends to steady-state within 30 min. The wax deposit mass is limited by thermal insulation of the cold surface and not wax depletion. As a function of rotational speed, the wax deposit mass is shown to be maximum under static condition and gradually decrease with increasing rotational speed (25 to 500 rpm). The wax deposit morphology is studied by cryo-scanning electron microscopy with the static deposit showing a layered structure (ridges) which diminished at 200 rpm, confirming a denser, less porous deposit. With its simple design and small sample volume, the CRF provides a route to easily study wax deposition under conditions more relevant to the industry scenario.

Introduction

Flow restriction in pipes by depositing wax on cold surfaces remains a significant problem in petroleum production, with the mechanism for wax deposition dependent on numerous physicochemical parameters. One parameter that is not always represented in lab-scale testing is fluid flow and its influence on wax deposition. Flow-through systems can provide such conditions but are often more complicated to use and rapid screening is not viable. A typical flow-through system consists of a feed vessel, pump and pipe-loop test section which is jacketed for coolant flow. Alternatively, the cold finger method has become an industry standard because of its small volume, good control of thermal conditions and ability to rapidly screen chemicals[1]. However, the method is often static and does not consider the effect of fluid hydrodynamics and shear on the wax deposit. To overcome this limitation there has been various adaptations to introduce fluid shear. Jenning and Weispefennig combined the cold finger and gentle fluid motion by stirring with a magnetic flea[2]. Higher shear stress and transition from laminar to turbulent flow has been achieved by mixing with impellers[3, 4]. While fluid flow is induced, the fluid stresses are highest at the impeller edges and not the depositing surface, the latter best representing the pipe flow environment.

Wu et al. developed a cold disk method to attain high shear stresses at the wax deposit-oil interface[5]. The cold disk was positioned in the outer cylinder wall, with fluid flow induced by a 2-paddle mixer. While the effect of flow velocity on wax deposit mass was studied, the flow pattern in the mixing vessel induced by a 2-paddle mixer is not known. Nigo et al. modified the cold-disc method and proposed a spinning-disc method to study fouling by fat deposits. The spinning-disc method provided a well-defined flow condition but only attained laminar flow and a shear stress that varied along the radius of the spinning-disc [6].

Zougari et al. developed an "Organic Solids Deposition and Control System" to study wax deposition under realistic production and transportation conditions. The method consists of a rotating cylinder submerged in a cold fluid chilled by an outer cylinder and could mimic production conditions in terms of Reynold number (*Re*) and wall shear stress. Wax deposition rates were shown to better represent data (comparison to the cold finger method) obtained using a pilot-scale flow loop[7]. However, the wax deposition is formed on the outer-stationary wall, in a region of low fluid shear stress.

Recognizing some test method limitations, the current study developed a cold finger method wherein the inner cylinder rotates (axially) and is chilled simultaneously. This set-up allows wax deposits to form in the region of high fluid shear stress under both laminar and turbulent flows.

CRF Design

Wax deposition was studied using the CRF apparatus shown in Fig. 1. The CRF consists of an inner rotating cylinder connected to a refrigeration unit (Grant LT ecoccol 100) via a rotary union coupling (Deublin 55-655-094), enabling restriction-free flow of the coolant through the rotating cylinder. A set of two-timing pulleys and a belt fastened the driveshaft to an overhead stirrer to precisely control the CRF between 0 and 500 rpm. For wax deposition, a cylindrical test sample (stainless steel 316L) was mounted on the inner rotating cylinder which was subsequently positioned in a doubled jacketed beaker used to regulate the temperature of the fluid using a heating bath (CW3-10 from Jeio tech, Inc. Lab companion). The dimensions of the cylindrical test sample and the jacketed beaker are provided in Table 1.

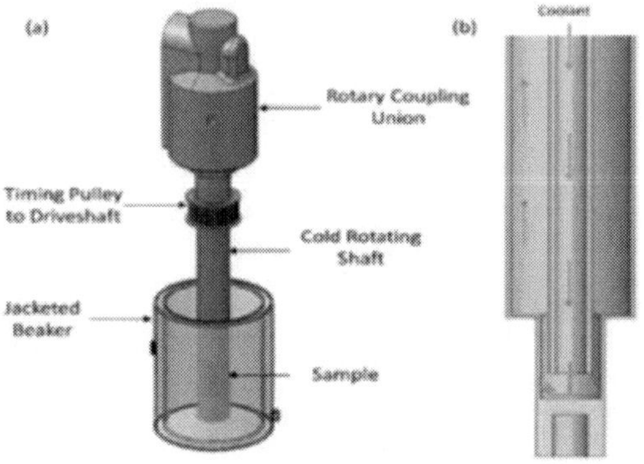

Figure 1. Schematic of the (a) CRF with the rotary coupling providing effective rotary seal and (b) coolant circulation path inside the rotating shaft.

Table 1. Dimensions of the cylindrical test sample and the jacketed beaker.

Cylindrical test sample	
Outer diameter	20 mm
Thickness	3 mm
Length	28 mm
Area	1759 mm^2
Jacketed beaker	
Inner diameter	66 mm
Volume	300 mL

The CRF flow field is Taylor-Couette-type (azimuthal flow between two cylinders) with a certain degree of axial circulation. While different from pipe flow, similarity between Taylor-Couette and pipe flow has been demonstrated[7-9]. The CRF is designed to provide a well-controlled thermal environment and achieve turbulent flow at low angular velocity. A turbulent flow (Re > 2000) is achieved when the rotational speed > 150 rpm. At the highest rotational speed (500 rpm) the wall shear stress is calculated to be 0.89 Pa, for a fluid viscosity of 0.00119 Pa.s (9 wt% paraffin wax-in-dodecane at 45°C). As the fluid viscosity increases towards the wax deposit, the wall shear stress will increase. While the wall shear stress may be lower than typically found in a production pipeline, the CRF provides insight into wax deposition kinetics under turbulent flow and increasing wall shear stresses.

Materials and Experimental Methods

Solid paraffin wax of low (C_{22}-C_{35}, melting point 53-57°C) and high (C_{25}-C_{45}, melting point 65-71°C) molecular weight with a purity > 97% and n-dodecane with a boiling point of 175-192°C (purity 96.5%) were purchased from Sigma Aldrich. All chemicals were used without further purification.

WAT measured by Crystal 16: The wax appearance temperature (WAT) of 9 wt% n-paraffin -in-dodecane (1:1 mass basis for low and high molecular weight wax) was measured by turbidity measurement performed at different cooling rates using a Crystal 16 apparatus (Technobis). The Crystal 16 comprises sixteen sample wells split into four blocks. Each block is independently temperature-controlled and enables a range of cooling rates to be studied simultaneously. Samples were added to 1 mL glass vials. The solution turbidity (transmission %) is measured as a function of temperature. Each test began by heating the samples to 60°C and held at temperature for 30 min to ensure thermal equilibrium. Then the samples were cooled at 0.5, 1, 2 and 3°C /min to 0°C while stirred using a magnetic flea. The WAT is taken to be the temperature at a light transmission of 90% (100% is transparent).

Wax deposition tests: Solutions of 9 wt% n-paraffin-in-dodecane were used for all tests. To prepare the wax solution, an equal amount of low and high molecular weight paraffin wax was added to n-dodecane at 60°C and held at temperature for 1 h until all the wax had dissolved (visual assessment). The solution was then cooled to 45°C to begin the wax deposition test. Deposition tests were conducted in a 300 mL Pyrex jacketed beaker which maintained the wax solution at 45°C (under static condition) and monitored using a K-type thermocouple (± 0.5°C). With the sample at 45°C, the coolant fluid at 5°C was pumped through the CRF to initiate wax deposition. For 0, 100, 200 and 500 rpm, wax deposits were collected at 1, 3, 5, 10, 15, 20, 25 and 30 min to study the kinetics of deposition. Dynamic conditions at 0, 25, 50, 100, 150, 200, 250, 300 and 500 rpm were compared at 30 min. The mass of the wax deposit on the cylindrical test sample was assessed by gravimetric measurement. Following each experiment, a scalpel was used to remove wax external to the cylindrical test sample. Mass deposited was taken to be the difference in mass between the cylindrical test sample in the absence and presence of wax. All wax deposition tests were run in triplicate.

Cryo-scanning electron microscope (Cryo-SEM): Cryo-SEM was used to study the morphology of wax deposits. Wax deposits were carefully removed from the cylindrical test sample using a clean scalpel. To avoid damaging the structure of the wax deposit, two vertical lines of ~10mm were cut from the edge of the sample and the scalpel slid between the sample and deposit surface to remove the wax layer. The wax layer was placed in a freezing rivet and the sample frozen in slushed nitrogen before transferring to the SEM under vacuum. To study cross-section of the wax layer, the sample was fractured using a chilled knife. The samples was sputter coated with Iridium and imaged using 2 kV at 0.1 nA.

Results and Discussion

Wax appearance temperature: Figure 2a shows data obtained from Crystal 16 for 9 wt% wax-in-dodecane cooled at rates between 0.5 and 3°C/min. The WAT is taken at the temperature when the transmission is 90% and Fig. 2b compares the WAT as a function of cooling rate. With increasing cooling rate the WAT was shown to decrease from 36.2°C to 32.0°C. The dependence of WAT on cooling rate is due to differences in crystal growth rate with temperature, with larger crystals formed at low cooling rates influencing solution transmission more than small crystals formed at high cooling rates.

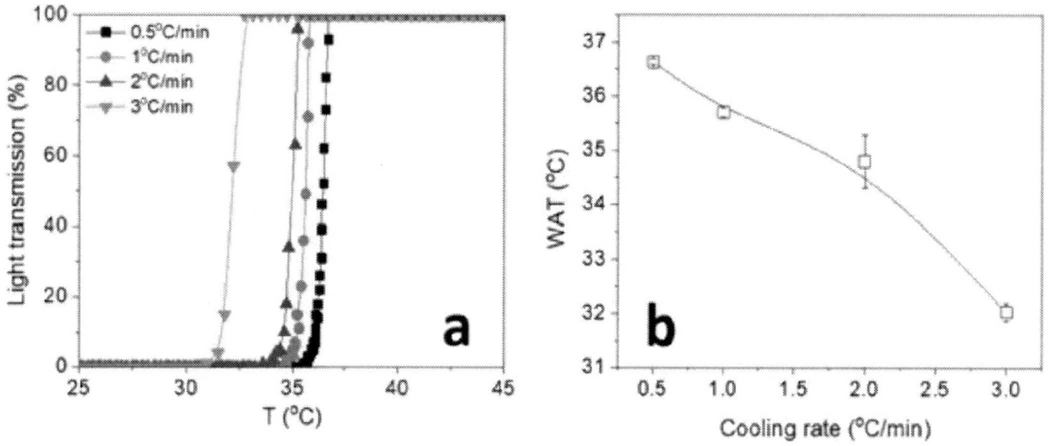

Figure 2. a) Light transmission % (turbidity) as a function of temperature for 9 wt% paraffin wax-in-dodecane during cooling from 60°C to 0°C at different cooling rates. The WAT is taken at the temperature when the light transmission is 90%. b) Comparison of WAT as a function of cooling rate.

Wax deposition kinetics: Figure 3 shows the wax deposit mass as a function of time and at rotational speeds of 0, 100, 200 and 500 rpm. For all tests, > 50% of the deposited wax (of the mass deposit at 30 min) occurred within 1 min. The initial rate of wax deposition is rapid and can be approximated by a linear rate. Due to the experimental method, data at t < 1 min could not be attained with good reproducibility. At t > 1 min the wax deposit rate is non-linear, and the deposited mass approaches a steady-state value at t = 30 min. Between t = 5 min and 30 min, the deposited mass increased by ~25%, ~18%, ~19% and ~25% for 0, 100, 200 and 500 rpm, respectively. Under static condition (0 rpm), the wax deposited at 24 h (4.9 g) confirmed the wax deposit continue to grow, although at negligible rates.

The wax deposition rate is attributed to the change in temperature at the wax deposit-oil interface. Based on the molecular diffusion mechanism, wax deposition depends on radial diffusion of wax molecules from the bulk fluid to the wax deposit-oil interface. Radial diffusion is induced by a temperature difference between the bulk oil (T > WAT) and the wax deposit-oil interface (T < WAT). WAT of 9wt% wax-in-dodecane is shown in Fig. 2b. While the temperature difference is initially high (reflecting the rapid increase in wax deposit), the temperature difference decreases with more wax deposited, thermally insulating the cold surface to the point when the temperature at the wax deposit-oil interface is equal to WAT and further wax deposition ceases[10]. Alternatively, wax deposition can cease if the wax in the bulk fluid is depleted. Based on the 9 wt% wax-in-dodecane, the initial mass of wax in solution is 22.5 g. For 0 rpm, the mass deposited at 30 min was 3.9 g, thus 83% of the initial wax content remains in solution and wax depletion cannot be attributed to the decrease in wax deposition rate.

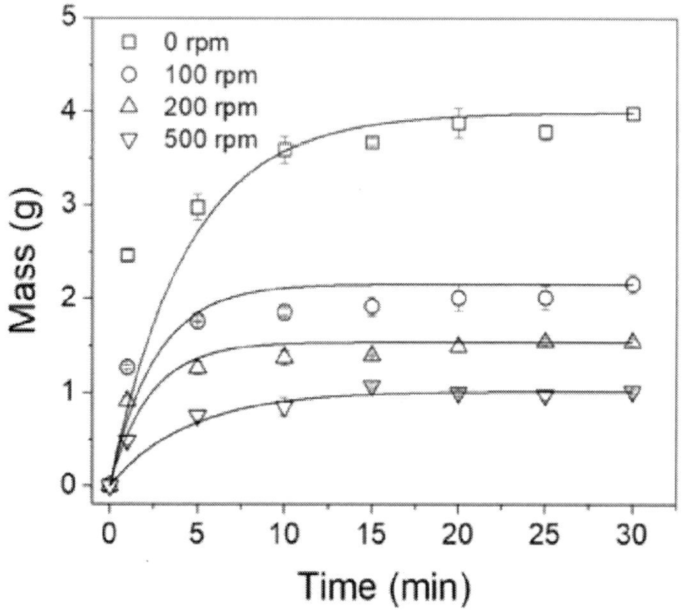

Figure 3. Time-dependent wax deposition as a function of CRF rotational speed. Experimental conditions: 9 wt% paraffin wax-in-dodecane, T_C = 5°C and T_F = 45°C. Lines are to guide the eye and takes the form, $y = y_0 + Ae^{(Bx)}$.

Wax deposition dynamics: Figure 4 compares the wax deposit mass at 30 min as a function of the CRF rotational speed between 0 and 500 rpm (the shear rate at 500 rpm is 750 s^{-1} based on a fluid viscosity of 0.00119 Pa.s, 9 wt% paraffin wax-in-dodecane at 45°C). The wax deposit mass decreases with increasing rotational speed, possibly attaining a minimum mass deposit at higher rotational speeds (where the mass deposit is independent of rotational speed), although higher rotational speeds have not yet been considered.

A decrease in wax deposit with increasing fluid flow is commonly observed and the reduction in deposit mass is found in both laminar and turbulent flows. The reduction in mass deposit is often considered to result from a decrease in the thermal gradient between the fluid and cold surface in flow and/or the effect of fluid shear stress on the wax deposit that removes wax [1, 11, 12].

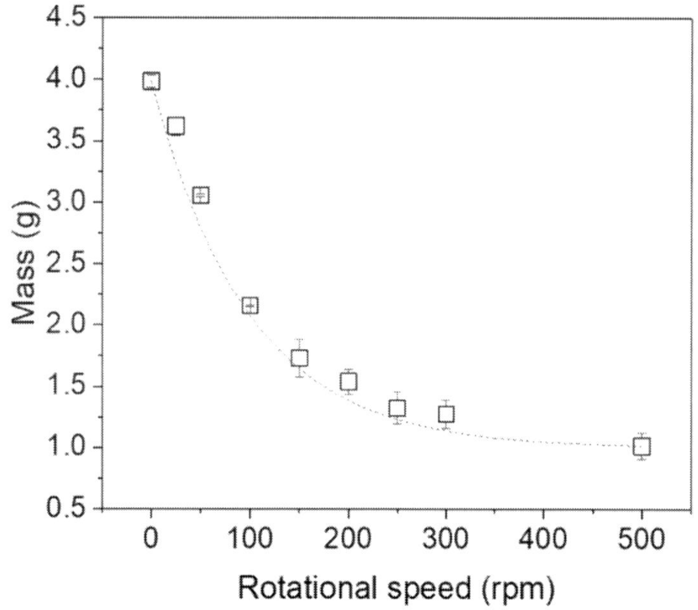

Figure 4. Mass of wax deposited at 30 min as a function of the CRF rotational speed. Experimental conditions: 9 wt% paraffin wax-in-dodecane, $T_C = 5°C$ and $T_F = 45°C$. Lines are to guide the eye and takes the form, $y = y_0 + Ae^{-\frac{x}{B}}$.

i) Heat transfer effect: An increase in rotational speed increases the heat transfer between the hot fluid and cold surface, thus the cold surface temperature increases and the bulk fluid temperature decreases, reducing the temperature difference. In the CRF, the temperature of n-dodecane in the absence and presence of 9 wt% paraffin wax was measured at different rotational speeds. With $T_c = 5°C$ and $T_F = 45°C$ (initial condition before CRF is submerged in the test fluid), 3 thermocouples positioned at the top, centre and bottom and at a radial distance of 10 mm from the cylindrical test sample, confirmed a reduction in fluid temperature with increasing rotational speed. For 0 rpm, T_F decreased to ~40°C (~10 min to attain steady-state values) and ~43.5°C (~10 min) in the absence and presence of wax. At 100 rpm, T_F decreased to ~36°C (~8 min, no wax) and ~42.5°C (~3 min, wax) and at 700 rpm, T_F decreased to ~34°C (~5 min, no wax) and ~ 42°C (~1 min, wax). The reduction in T_F will contribute to an increase in T_C due to forced convection from the hot fluid to cold surface. Unfortunately, T_C could not be directly measured due to challenges associated with measuring temperature on a rotating surface. However, the increase in T_C with a decrease in T_F was verified at 0 rpm when T_F reduced to 40°C after 12 min, due to natural convection, and led to an increase in T_C to 15°C. The change in heat transfer due to the rotational speed is not linear, with a significant reduction between 0 and 100 rpm and then an almost independence at ≥ 500 rpm. It is worth noting that forced convection will influence T_F and T_C more than natural convection. At rpm ≥ 500, the decrease in wax deposit mass is not due heat transfer (since T_F and T_C are less independent of rpm) and may indicate that the further reduction in wax deposit mass at 500 rpm compared to 300 rpm likely results from another factor.

ii) Shear removal: Shear removal describes either the retardation of continuous wax deposition (wax molecules adhere less on the wax deposit) or sloughing (the removal of wax from the wax deposit). Sloughing is thought to occur when the fluid shear stress exceeds the wax yield stress causing deformation of the wax layer and ablation of wax from the deposited layer.

For the CRF, sloughing as a function of rotational speed was studied on wax deposits formed under static conditions for 30 min (the wax deposits were thick and retained ~85 vol% solvent). After wax deposition, the CRF was submerged in pure n-dodecane (no wax dissolved) and the wax deposit rotated at 100, 200, 300 and 500 rpm for 10 min. To minimize dissolution of wax T_F was set to 32°C (below the WAT). GC-FID analysis of the n-dodecane after the CRF was submerged in pure dodecane for 30 min (no rotation) revealed only trace amounts of C20-C25, and therefore the contribution of dissolution to changing wax deposit mass is considered negligible.

Figure 5 shows the wax deposit after rotating the cylindrical test sample for 10 min at increasing rotational speeds. While the wax deposit at 100 rpm shows slight deformation of the deposit (following the standard protocol), the wax deposit mass was within ~95% of the static wax deposit, hence no measurable mass loss. With increasing rotational speed, the wax deposit is visually deformed, and undulations were observed throughout the wax deposit at 500 rpm, with the mass loss increasing by ~15%, ~20% and ~30% for 200, 300 and 500 rpm, respectively.

While these results confirm the potential of sloughing, it should be noted that these measurements were all conducted on wax deposits formed at 0 rpm. It is not clear if sloughing is a contributing factor when depositing at higher rotational speeds. While the wax deposit layer is thinner at higher rotational speeds, Fig. 6b also confirms a different in wax deposit structure which may be more resistant to sloughing.

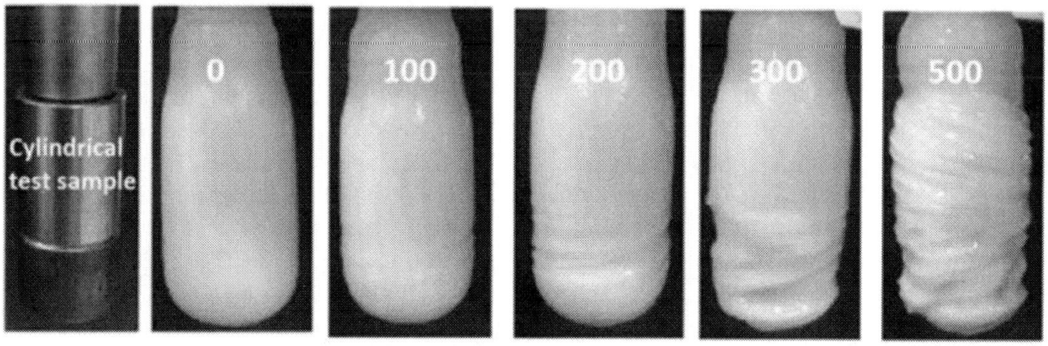

Figure 5. 9 wt% paraffin wax-in-dodecane deposited on the CRF at 0 rpm for 30 min. With the wax deposit formed, the sample is transferred to pure dodecane and rotated at 100, 200, 300 and 500 rpm for 10 min.

Wax deposit structure: Cryo-SEM images of wax deposits formed under static (0 rpm) and dynamic (200 rpm) conditions are shown in Fig. 6. For static deposition (Fig. 6a), the wax deposit reveals an arrangement of thin sheets of paraffin stacked together in parallel multi-layers with space in between where the solvent is occluded in the deposit. The layer can be described as a semi-solid (composite of paraffin wax and oil), with stepped morphology of wax layers that result in ridges throughout the wax deposit. At 200 rpm, the stepped morphology of wax layers is less visible with negligible porosity throughout the wax layer (based on the uniformity of the wax layer), indicating a denser wax deposit. In the absence of shear, paraffin crystallizes in layers which consist of wax molecules having planar zigzag conformations, and this results in the stepped morphology. In shear, wax crystals are found to be smaller and have a tendency to aggregate, which in the bulk fluid can lead to stronger networks[13, 14]. In Fig. 6b the wax deposit is visibly different to that formed under static condition and shows reduced stepped morphology. The different morphology may be influenced by the fluid shear during wax deposition and formation (rearrangement of wax crystals), and/or densification of the wax deposit layer through continued shear.

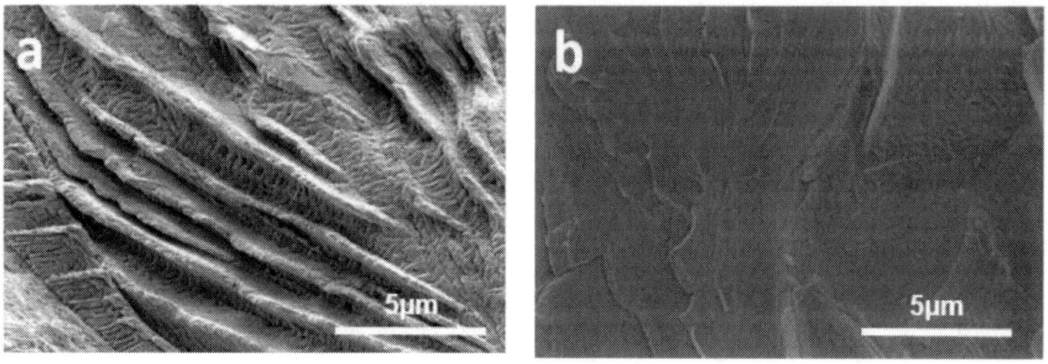

Figure 6. Cryo-SEM images of wax deposits formed after 30 min under static (a) and dynamic (b) conditions.

Conclusion

A novel Cold Rotating Finger representing a Taylor-Couette flow has been developed to study wax deposition. Both kinetic and dynamic studies were considered and the kinetic study showed an initial high deposition rate (within 1 min) followed by a secondary slower deposition rate as the wax deposits approached steady-state values (~30 min). Wax deposition kinetics are shown to be governed by the temperature gradient between the bulk fluid and the wax-oil interface. Initially, the temperature gradient is large and promotes a high deposition rate of wax. As the wax deposit thickness increases the temperature gradient decreases and this results from the increased thermal resistance of the wax deposit layer. The wax deposit approaches a steady-state value (deposit mass) as the wax-oil interface temperature approaches the WAT. Dynamic studies showed a decrease in the wax deposit mass with increasing rotational speed of the CRF. While the reduction in mass can be attributed to a heat transfer and/or shear removal effect, in the current study we attribute the reduction in wax deposit mass to a predominantly heat transfer effect. Increasing the rotational speed decreases T_F and therefore increases T_C, thus decreasing the thermal diffusivity of wax molecules. While this phenomena describes behaviour at low rotational speeds, at 500 rpm the fluid shear was shown to strongly deform and decrease the deposit mass of a wax layer formed at 0 rpm. This introduces the possibility of shear removal (sloughing) at higher rpm, and it may be justified since the steady-state value at 500 rpm appears to be lower than the projected trend for steady-state values at 300 rpm and less.

References

1. Huang, Z., S. Zheng, and H.S. Fogler, *Wax Deposition*. 1 ed. Emerging Trends and Technologies in Petroleum Engineering. 2016: CRC Press.
2. Jennings, D.W. and K. Weispfennig, *Effects of shear and temperature on wax deposition: Coldfinger investigation with a Gulf of Mexico crude oil.* Energy and Fuels, 2005. **19**(4): p. 1376-1386.
3. Dos Santos, J.D.S.T., A.C. Fernandes, and M. Giulietti, *Study of the paraffin deposit formation using the cold finger methodology for Brazilian crude oils.* Journal of Petroleum Science and Engineering, 2004. **45**(1-2): p. 47-60.
4. Kasumu, A.S. and A.K. Mehrotra, *Solids deposition from wax-solvent-water "waxy" mixtures using a cold finger apparatus.* Energy and Fuels, 2015. **29**(2): p. 501-511.
5. Wu, C.H., et al., *Measurement of wax deposition in paraffin solutions.* AIChE Journal, 2002. **48**(9): p. 2107-2110.

6. Nigo, R.Y., et al., *Experimental studies of freezing fouling of model food fat solutions using a novel spinning disc apparatus.* Energy and Fuels, 2009. **23**(12): p. 6131-6145.
7. Zougari, M., et al., *Novel organic solids deposition and control device for live-oils: Design and applications.* Energy & Fuels, 2006. **20**(4): p. 1656-1663.
8. Akbarzadeh, K. and M. Zougari, *Introduction to a novel approach for modeling wax deposition in fluid flows. 1. Taylor-couette system.* Industrial and Engineering Chemistry Research, 2008. **47**(3): p. 953-963.
9. Zougari, M.I., *Shear driven crude oil wax deposition evaluation.* Journal of Petroleum Science and Engineering, 2010. **70**(1): p. 28-34.
10. Wang, W. and Q. Huang, *Prediction for wax deposition in oil pipelines validated by field pigging.* Journal Of The Energy Institute, 2014. **87**(3): p. 196-207.
11. Huang, Z., et al., *The Effect of Operating Temperatures on Wax Deposition.* Energy & Fuels, 2011. **25**(11): p. 5180-5188-5180-5188.
12. Lu, Y., et al., *Counterintuitive effects of the oil flow rate on wax deposition.* Energy and Fuels, 2012. **26**(7): p. 4091-4097.
13. Blake, A. and A. Marangoni, *The Use of Cooling Rate to Engineer the Microstructure and Oil Binding Capacity of Wax Crystal Networks.* Food Biophysics, 2015. **10**(4): p. 456-465.
14. Blake, A. and A. Marangoni, *The Effect of Shear on the Microstructure and Oil Binding Capacity of Wax Crystal Networks.* Food Biophysics, 2015. **10**(4): p. 403-415.

572239 Use of Thermal Analysis and Rheometry to Study the Properties of Crude Oil

Crude oil is typically extracted as a single organic liquid phase from the reservoir. However during transportation, a drop in the temperature can cause the higher molecular weight paraffins or waxes to separate into the solid phase. The temperature at which the first wax crystals appear during cooling is commonly known as the wax appearance temperature (WAT). Once below the WAT, the crude oil can form a gel-like structure. Breaking down this structure to restart and maintain flow may often require energy- and cost-intensive approaches.

This talk will focus on characterizing crude oil and related fuels using thermal and rheological analysis at different temperatures above and below the WAT in the context of flow assurance. Specifically, we will compare four different types of light crude oils for differences in their properties. The thermal analysis section will emphasize on use Modulated DSC ® (MDSC) as a technique to measure WAT with improved sensitivity for analyzing light crudes with the additional benefit of reduced supercooling. The rheometric studies will cover a range of studies from viscosity measurements above the WAT to measuring the yield stress below the WAT. Given that WAT, viscosity and yield stress play a critical role in flow assurance studies, the emphasis of this talk will be on the method development to perform these measurements along with strategies to eliminate or reduce experimental artefacts.

Note: Modulated DSC ® is a registered trademark of TA Instruments.

577943 Asphaltene Destabilization Mechanism Investigated By Ultra-Small Angle X-Ray Scattering and *in Situ* Titration

Asphaltene destabilization causes plugging in oil-transportation pipelines and induces complications in refineries by fouling. Micron size insoluble asphaltenes clusters have been extensively investigated by optical microscopy (>1 micron detection), and soluble asphaltene behavior has been studied by small angle X-ray scattering (<50 nm detection). However, the gap in characterization from 50 nm to 1 micron has left the mechanism of asphaltene destabilization obscure. Ultra-small-angle X-ray scattering (USAXS) was recently demonstrated by our team to be a powerful technique that could monitor this previously overlooked length scale between 50 nm and 1 micron (Yang et al., 2018) (Ismail et al., 2018). However, the rate of asphaltene destabilization can vary significantly with slight changes in heptane concentration around the "onset point," and *in situ* monitoring of asphaltene destabilization is necessary to capture the initial stages of this process. In this talk, we will discuss the asphaltene destabilization mechanism using continuous length scale monitoring to understand the gradual association of asphaltene molecules to micron size insoluble clusters. We implemented a flow-cell system to investigate the real-time formation of nanoaggregates, soluble and insoluble clusters. Heptane was used as an anti-solvent and titrated into an asphaltenes-toluene solution at a different flow rate for precise control of the rate of asphaltene destabilization. USAXS was used to monitor the overall destabilization process at a time resolution of <5 minutes. This technique is effective to probe the size, internal structure, overall shape, and surface morphology within a range from 1 nm to 5 microns. The results demonstrate that asphaltenes aggregate together to form insoluble particles with size on the order of 100 nm which then further cluster form insoluble fractal flocs of size >1 micron. Direct monitoring of the asphaltene destabilization process from the initial stages of aggregation is essential to understand the best modeling approaches for phase behavior, aggregation, deposition, and inhibition mechanisms of action. These results help to understand the role of solvent environment, temperature, and pressure on asphaltene stability to provide constructive knowledge for asphaltenes inhibitor research.

572846 Invited: Sensor Applications in Bioengineering: Detection of Small Molecules within Complex and Three-Dimensional Biological Milieu

Bioengineering applications in regenerative medicine and single-cell analysis present challenging microenvironments for diagnostic and real-time detection. We have developed and applied sensors to detect small molecules in neuronal, cardiovascular, blood and biofilm systems. In particular, we have aimed to establish easy-to-use sensor platforms that do not interfere with biological processes. These sensors are reagentless and can be deployed in the presence of complex macromolecular solutions that readily foul, degrade or block many traditional sensing approaches. Herein, we describe our efforts towards detection of neuroactive molecules in the brain, hypoxia gradients in three-dimensional cellular scaffolds, and subcellular oxygen concentrations during electrophysiological and chemo-mechanical single cell experiments.

575136 Tissue-like Neurochemical Sensors Operating in Brain and Gut

Normal neurotransmission relies on a precise and delicate balance of chemical signaling mediated by the release of variety neurotransmitters. Among them, the monoamines including dopamine and serotonin are uniquely involved in the regulation of cognitive processes such as emotion, arousal and memory. Therefore, it is of great interest in both basic research and clinical studies to monitor the fluctuation of monoamine in both brain and peripheral nervous systems. As a very soft and complex biomaterial, the human brain is under constant motion and deformation because of different physiological dynamics, while GI tract is a series of soft organs with inherent motility, joined in a long and twisting tube from the mouth to the anus. As a result, rigid neurochemical sensors can cause inflammation to the brain tissue, and neurochemical sensing in awake animals have historically been impossible to traditional rigid and fragile devices.

Softness and stretchability in new-generation neuroengineering tools are essential to minimize the perturbation of motions inherent in biological systems and maximize the physiological fidelity. Here we develop a new neuroengineering platform, based on soft materials, that could allow access to the brain and entire enteric nervous system for neurochemical sensing. The new neurochemical sensor is based on graphene-elastomer composite, which offers chronic, sensitive, and selective sensing of dopamine and serotonin in the brain. Ex vivo studies shows that the soft sensor could accommodate the peristalsis movement of the intestine and continuously monitor the serotonin dynamics via thermal treatment. In vivo experiments also show that the sensor could be implanted in the colon of mice to examine the drug-stimulated serotonin response. The new neuroengineering platform described here can provide unprecedented insights on the microbiota–gut–brain communication in the context of mental disorders and therapeutic interventions, and can lead to new therapies for psychiatric disorders through GI tract without invasive surgery to the brain.

570825 Cell-Based Electrochemical Sensors for Environmental Pollutant Detection

Endocrine disrupting compounds are ubiquitous; they are found in many products intended for common use, including pesticides, plastics, and even some medications. Yet, these chemicals are implicated in health problems ranging from diabetes to cancer. We have developed a cell-based electrochemical strategy to rapidly and sensitively quantify these compounds. Our strategy measures the biological activity of chemically dissimilar xenoestrogens by harnessing the specificity of a human estrogen receptor. This receptor is expressed on the surface of *E. coli* as a component of an electrochemical sandwich assay. Our sensors require no specialized skills to implement and have enabled very sensitive detection (sub-ppb levels) of environmental pollutants. We have successfully detected compounds in from complex solutions, such as baby formula, and have measured estrogenic activity from unknown compounds leeched from a BPA-free baby bottle.

577305 Biomimetic Composites As Microfabricated, Flexible, Degradable Electrochemical Sensors

Biomimetic composites as microfabricated, flexible, degradable electrochemical sensors

- M. Xu, S. Pradhan, R.K. Pal, V.K. Yadavalli

Department of Chemical and Life Science Engineering, Virginia Commonwealth University, Richmond, VA

Biomimetic composites of naturally derived and synthetic polymers provide exciting opportunities to develop mechanically flexible, physiologically compliant, and degradable bioelectronics systems. In prior work, we have shown that naturally derived silk proteins and conducting polymers (specifically poly(3,4-ethylenedioxythiophene) :poly(styrene sulfonate) (PEDOT:PSS)) can form functional bioinks for sensing and other electrochemical applications. These inks combine the mechanical strength, biocompatibility and degradability of silk proteins, with the conductivity and chemical functionalization of the PEDOT:PSS. The fabrication and characterization of photopatternable, water-based conductive inks comprising PEDOT:PSS and synthesized photoreactive silk proteins was earlier shown [1] The presence of photoreactive groups permits a fully aqueous photolithographic strategy to form conductive microelectrodes on both rigid substrates as well as flexible silk films [2].

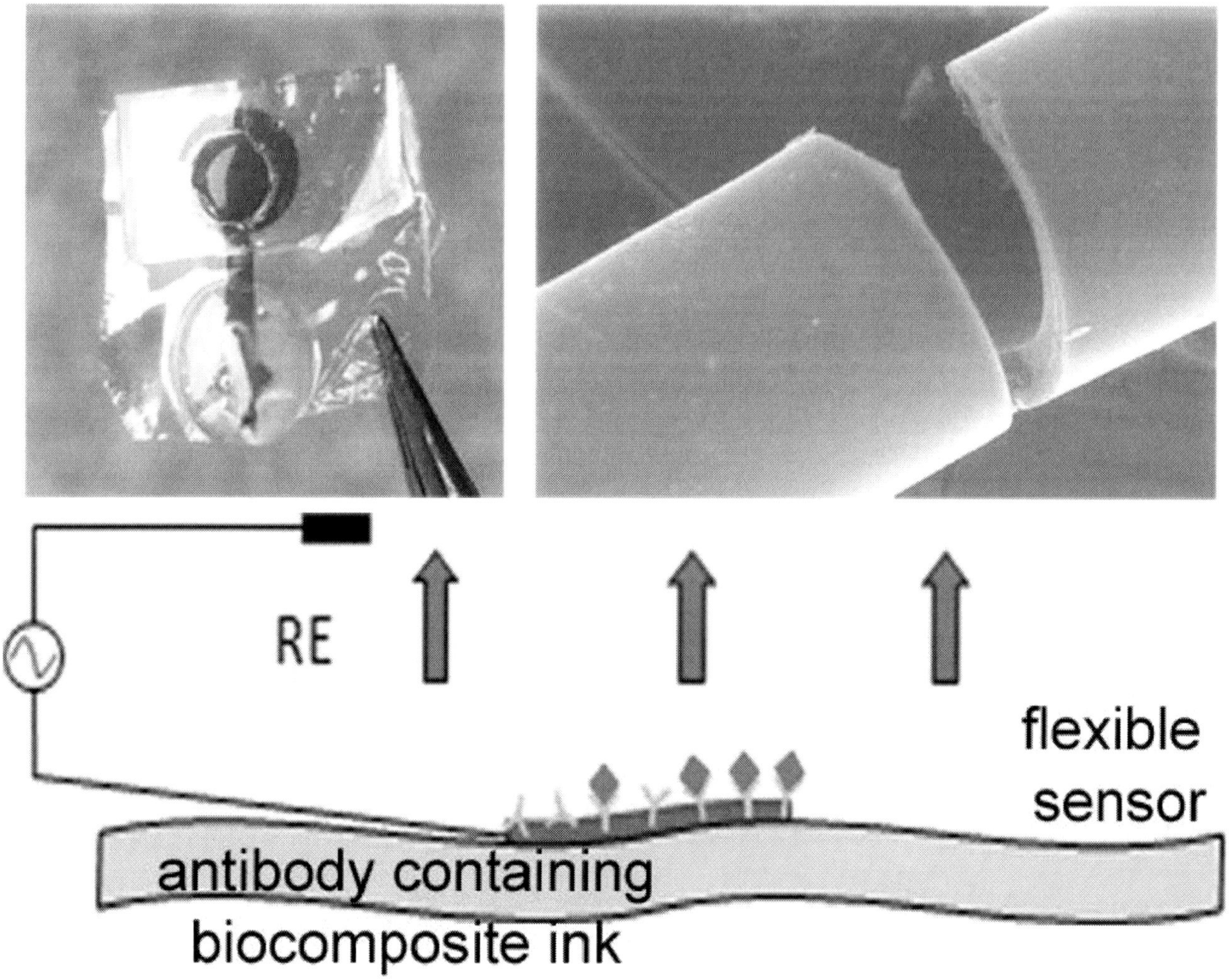

Here we will present the latest developments using this technology, showing how these devices can function for flexible bioelectronics applications in the form of microelectrodes for ultrasensitive detection, as well as components such as microsupercapacitors.[3, 4] We show how electroactive biomolecules such as neurotransmitters can be detected sensitively, while non-electroactive biomolecules such as glucose and glutamic acid, can be detected by encapsulating specific enzymes. Due to the room temperature, aqueous fabrication process, we can immobilize antibodies in these sensors resulting in flexible immunosensors for targets such as vascular endothelial growth factor (VEGF) for the first time [3]. The electroactivity of conductive ink can be improved by the addition of small amounts dopants. Using these doped composites, we further demonstrate flexible energy storage devices due to the capacitive nature of the biomaterial. The presence of silk proteins as the matrix of the composite makes it completely biodegradable and biocompatible, potentially resulting in transient or implantable devices. This platform can also be used for the specific control of cell culture in flexible sheets. The mechanical, biochemical and electrochemical characterization of the composite and its microfabrication are discussed. By virtue of a range of versatile properties, utility as bio-sensors, opto-electronic devices and flexible energy storage systems are envisioned.

References:

[1] "Photolithographic micropatterning of conducting polymers on flexible silk matrices" - RK Pal, AA Farghaly, MM Collinson, SC Kundu, VK Yadavalli, *Advanced Materials*, 28(7), 1406-1412, 2016

[2] "Conducting polymer-silk biocomposites for flexible and biodegradable electrochemical sensors" - RK Pal, AA Farghaly, MM Collinson, SC Kundu, VK Yadavalli, *Biosensors & Bioelectronics*, 81, 294-302, 2016

[3] "Flexible biosensors for the impedimetric detection of protein targets using silk-conductive polymer biocomposites" – M Xu, VK Yadavalli, *ACS Sensors*, in press, 2019

[4] "Fabrication of Flexible, Fully Organic, Degradable Energy Storage Devices Using Silk Proteins" - RK Pal, SC Kundu, VK Yadavalli, *ACS Applied Materials & Interfaces* 10 (11), 9620-9628, 2018

569635 Nanopore Based Highly Sensitive and Accurate Detection of HIV

Nanopore Based High Performance Point-of-Care Biosensor for Acute HIV Detection

Xiaojun Wei, Chang Liu

Human immunodeficiency virus (HIV) is the causative agent of acquired immuno-deficiency syndrome (AIDS). There are over 1.8 million infections HIV occur annually worldwide, and 30%¨C50% are attributed to recently infected individuals that are unaware of their disease state. Rapid diagnosis of acute HIV (within 3 weeks of infection) may reduce new cases by 10%¨C25%. The fourth-generation rapid diagnostic test for HIV is currently employed worldwide as the frontline screening test. This Enzyme-linked Immunosorbent Assay (ELISA) based test is designed to detect HIV-1 p24 antigen in addition to HIV antibodies to achieve early detection. However, several recent field tests have indicated unacceptably low diagnostic sensitivity (0%) of acute HIV-1.

Herein, a nanopore biosensing platform with single-molecule level sensing and analysis capacity has been developed to address the unmet need for rapid and accurate diagnosis of acute HIV to reduce transmission. In our approach, p24 antigens isolated from plasma samples were cleaved into several specific peptide sequences, which can be recognized and quantified by their signature signals during translocation through the nanopore. A portable QuantiPep-HIV prototype suitable for point-of-care settings was developed and tested with serum samples from HIV-infected and uninfected individuals to demonstrate the feasibility and to evaluate the diagnostic performance. Our results showed that QuantiPep-HIV can improve the diagnostic sensitivity and specificity of acute HIV.

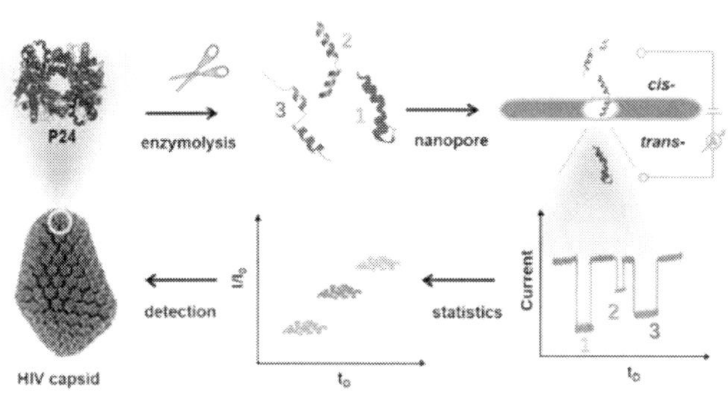

Figure. Schematic diagram for detecting peptide fragments from p24 antigens by nanopore

582378 Wireless, Soft, Skin-integrated Microfluidic Systems for Electrochemical/Colorimetric Biosensing and Capture of Sweat

Human eccrine sweat is a rich composition of nearly 60 known biochemicals including metabolites, electrolytes, vitamins, amino acids, exogeneous drugs, and small proteins. Advances in sensing technologies, conformal electronics, and materials science lay the foundation for recent demonstrations of wearable sweat sensors. Unfortunately, majority of the devices rely either on active, battery-powered electronics for electrochemical detection or passive, colorimetric chemistries for visual readout. Complex construction, large size/weight and possibility of battery leakage represent disadvantages of the former; semi-quantitative operation and limited scope of measurable biomarkers are drawbacks of the latter. Additionally, present wearable sensors can detect only a handful of biomarkers present in the sweat. Quantification of the remaining chemicals still requires sweat collection using adsorbent pads followed by off-site chemical analysis. Such a route provides average concentration of sweat analytes and restricts the acquisition of their temporal concentrations. This prohibits the understanding of the relation between dynamic variations in sweat chemistry and human physiology. In this talk, I will describe an unusual collection of enabling materials, sensing technologies, advanced microfluidics, stimulus dependent biocompatible power systems, galvanic time-tracking schemes, and wireless electronics to address these issues. Specifically, I will describe my strategies that include soft microfluidics, fuel cells, trigger-activated biocompatible batteries, colorimetrics and near field communication technologies that result in new classes of wireless, skin-integrated devices for detailed analysis of human sweat.

574301 The Electrochemical Impedance Response of a Continuous Glucose Monitor

The Gerischer impedance has been observed in systems with an electrochemical reaction coupled with a homogeneous chemical reaction. Under the assumptions of linear homogeneous reaction, equal diffusion coefficients and a semi-infinite condition, the Gerischer impedance can be solved analytically [1, 2]. However, with non-ideal electrochemical systems, where coupled nonlinear homogeneous reactions and various mass-transfer properties make exact analytical resolution impossible, a more detailed analysis is required.

In the present work, a mathematical model for the impedance response of glucose-oxidase based electrochemical biosensors has been developed [3]. The complicated coupling between thirteen homogeneous reactions and three heterogeneous reactions has been considered in the model. The homogeneous reactions included anomerization between α-D-glucose and β-D-glucose and four reversible enzymatic catalytic reactions transforming β-D-glucose and oxygen into gluconic acid and hydrogen peroxide, pH-dependent enzymatic activity and a biological buffer system. The electroactive hydrogen peroxide was considered to be reversibly oxidized or reduced at the electrode.

The mathematical model is solved numerically by using the finite-difference method and Newman's BAND algorithm [4]. The model demonstrates how the coupled non-linear homogeneous reactions effect the diffusion impedance, which has broadened the scope of the Gerischer impedance. The model can be used to explore the influence of various system parameters on limiting current, reaction profiles, and diffusion impedance. The system parameters, including interstitial glucose concentration, oxygen concentration, active enzyme concentration, diffusion coefficients, reaction rate constants and layer thickness, are related to various sensor working conditions such as body sugar level, inflammation, sensor degradation and sensor design.

References

1. H. Gerischer, "Wechselstrompolarisation Von Elektroden Mit Einem Potentialbes- timmenden Schritt Beim Gleichgewichtspotential," *Zeitschrift fur Physikalische Chemie*, **198** (1951) 286–313.
2. M. E. Orazem and B. Tribollet, *Electrochemical Impedance Spectroscopy* (John Wiley & Sons, Hoboken, NJ, 2017), 2nd edition, p. 279-293.
3. M. Gao, M. S. Hazelbaker, R. Kong, and M. E. Orazem, "Mathematical Model for the Electrochemical Impedance Response of a Continuous Glucose Monitor," *Electrochimica Acta*, **275**(2018),119-132
4. J. S. Newman and K. E. Thomas-Alyea, *Electrochemical Systems* (John Wiley & Sons, Hoboken, NJ, 2004), 3rd edition.

Acknowledgement

The support of Medtronic Diabetes (Northridge, CA) and Andrea Varsavsky, program monitor, is gratefully acknowledged.

564028 Converting Low-Pressure Natural Gas to Foamed Stimulation Fluid: Defining the Operating Envelope

As part of a Department of Energy (DOE) sponsored program, an optimal thermodynamic pathway to transform natural gas (NG) into pressurized NG suitable for use as the internal phase in a foamed fracturing fluid has been already identified. Use of NG foamed fracturing fluids has the potential to reduce the amount of water requirements for stimulation and pose benefits for productivity in water-sensitive formations. This study aims to extend the investigation to determine key reservoir features where using a NG foam fracturing fluid may be advantageous. We provide a comparative analysis, using validated models, to determine the advantages of using natural gas foams relative to conventionally used slickwater, linear gel and crosslinked fluid.

A full 3D reservoir model in a shallow Fayetteville Shale formation has been constructed. Fundamental laboratory and pilot field tests data have been collected for NG foam fluid properties for numerical modeling. Rheology, friction and leak-off properties of the fracturing fluids have been incorporated in creating a numerical model. A 3D-complex hydraulic fracture simulation model incorporating 1D and 2D particle transport models have been utilized. Numerical reservoir simulation of over 50 different sensitivity scenarios has been incorporated for fracture modeling as well as gas production evaluation. Owing to lower density than conventional liquid column, NG form fluid are likely to result in higher surface pressure. Therefore, NG foamed fluid for deeper sections of the Fayetteville formations has been modeled at various degrees of pressure loss per feet to determine the operational depth limits of NG fracturing fluid application in the area.

Reduced pump rate with NG foamed fracturing fluid leads to a lower frictional pressure loss in tubing without compromising the ability to place the desired amount of proppant in the formation. In recent flow loop experiments, non-Newtonian shear-thinning NG foamed fracturing fluid exhibits a higher effective viscosity that allows for effective transport of the proppant. Modeling results indicate that proppant placement is at least 10% farther in the hydraulic fractures than with linear gel and slickwater. Economic evaluation of NG foamed fracturing operations reveal that NG foam fracturing fluid improves economic returns by over 12% as it lowers the water handling, transportation and disposal costs for the operator while achieving superior productivity. Water sensitive and clay rich layers have shown greater production potential with NG based foam fracturing fluid than slickwater, linear gel and crosslinked fluids.

Although foamed fluids were first reported to be used in 1960s the usage of nitrogen and carbon dioxide foams has not been widely practiced due to cost, complexity, and unproven production benefits. Use of NG foamed fracturing fluid is not widespread either, but this study attempts to identify specific regions and reservoirs where use of these fluids may lead to economic and long-term production benefits. We believe that use of natural gas foams is likely to provide long-term sustainable benefits in areas where water procurement and disposal costs are high, where natural gas may be available from a central processing facility via pipelines and where the reservoir is relatively shallow and contains clay-bearing minerals.

564461 Impact of Zirconia Nanoparticles and Polymer on Drilling Fluid Rheology, Fluid Loss and Filter Cake Properties

Nanotechnology has been widely employed in the oil gas sector to improve the extraction of fossil fuels from reservoirs. Nanoparticles in the formulation of drilling fluids showed remarkable properties especially reduced fluid loss and filter cake thickness, better rheological properties and stabilized reactive shale formations. This paper discusses the use of zirconia nanoparticles and high molecular weight polymer for rheology, fluid loss and filter cake properties. The shale inhibition test was performed to study the effect of nanoparticles, polymer hydration, and swelling behavior of shale.

The water-based drilling fluid was prepared by mixing the different concentrations of polymers and nanoparticles along with other additives such as polyanionic cellulose, sodium silicate, sodium hydroxide as pH control agent and barite. Numerous experimental techniques were employed such as linear swelling test, hot rolling dispersion test, low-pressure/low-temperature, and high-pressure/high-temperature filtration tests, zeta potential, shale surface hydration test, and shale compressive strength test. The filtration and rheological properties of drilling fluid were determined to various temperatures ranging from 75°F to 400°F.

The different concentrations of zirconia nanoparticles and polymer were used to formulate the modified drilling fluids and results were compared with the conventionally used drilling fluid. The rheological properties of modified drilling fluid increased by increasing the concentration of nanoparticles in the drilling fluid. The fluid loss volume of modified drilling fluid was reduced by 24% compared to the fluid loss volume with the base fluid. The filter cake thickness has been reduced significantly by adding nanoparticles in the drilling fluid. The shale swelling and hot rolling dispersion tests with modified drilling fluid showed that swelling of shale decreased prominently compared to the base drilling fluid. The synergistic effect of high molecular weight polymer and zirconia nanoparticles protects the surface of shale with minimum swelling and hydration. The rheological properties of drilling fluid at 400 °F was slightly affected. The fluid loss results showed promising results with the incorporation of zirconia nanoparticles in drilling fluid. The optimum nanoparticle concentration was determined at 0.1% that showed best rheological and fluid loss properties.

The use of zirconia nanoparticles along with high molecular weight polymer enhanced the rheological and filtration properties of drilling fluids for high temperature applications. The shale hydration and swelling of shale significantly reduced by utilizing the nanoparticles in the drilling fluids. The optimum concentration of 0.1% of nanoparticles were selected for enhanced properties of drilling fluids.

572097 Brine-Soluble Zwitterionic Polymers with Tuneable Adsorption on Carbonates

Injection of aqueous fluids into reservoirs for enhanced oil recovery (EOR) has been of great interest in petroleum engineering. EOR using viscous polymer solutions increases the fluid viscosity and reduces its relative permeability in the reservoir, therefore improving volumetric sweep efficiency for EOR and oil recovery. However, significant polymer adsorption on the rock surface and in the pores reduces fluid viscosity and increases the overall cost of EOR as more materials are required to compensate for the lost polymers. Polymer retention in high salinity and high temperature reservoir conditions is one of the greatest challenges in polymer flooding EOR. Here we report the dynamic adsorption and retention studies of five brine-soluble zwitterionic polymers using a Quartz Crystal Microbalance (QCM-D) and core flooding at high temperatures. Our results indicate that the dynamic adsorption of the polymers on carbonates correlates to their structure variation. QCM-D coupled with core flooding experiments help to understand the dynamic adsorption behavior of these polymers with clear insight on the mechanisms and the kinetics of adsorption. This work presents a unique way of correlating the structural variations of polymers to the dynamic retention on carbonate surfaces under extreme reservoir conditions.

584596 Green Solvent for Enhanced Recovery in Unconventional Resources

In the past decade, the oil and gas industry has made tremendous improvements in horizontal drilling and well stimulation techniques, yet recovery rates for unconventional oil and gas reservoirs continue to remain quite low. Typically, less than 10% of the original oil in place (OOIP) is recovered. A key challenge for oil recovery in unconventional formations is accessing the oil that resides in the low-permeability, low-porosity rocks. While induced fracture networks certainly help to connect pore networks and create pathways to enable oil production, significant volumes of oil remain trapped in the unfractured reservoir matrix.

In certain reservoirs, such as the Bakken and Eagle Ford oil plays, pore spaces are clogged by immobile organic phases that impede the flow of oil and gas. The immobile phases are typically separated into two classes based on solubility: kerogen (the insoluble component) and bitumen (a soluble component in standard organic solvents). Kerogen exists as a solid and hosts the nanoscale pore network partially responsible for hydrocarbon storage and transport. Bitumen exists as a semi-solid or viscous fluid similar to shoe polish with a high viscosity that renders it essentially immobile. This viscous (but soluble) organic phase clogs pores, swells kerogen and ultimately, impedes production from certain formations where bitumen is abundant.

Bitumen is relatively abundant in lower thermal maturity shales (i.e. shales that have not been subjected to much heat in the subsurface), and the loss of oil recovery due to bitumen can be dramatic. For example, the Upper and Lower Bakken intervals have relatively high volumes of OOIP; however, these zones have so much bitumen that none of the oil can be economically recovered. Instead, essentially all the production from the Bakken system comes from the Middle Bakken and Three Forks intervals, which have lower volumes of OOIP but also much lower bitumen content. The same is true in the Eagle Ford; many operators produce from zones of low bitumen content, but zones that are richer in bitumen are not produced.

One potential EOR method to produce these bitumen-rich zones is to apply a chemical treatment that removes the soluble bitumen from the formation. In laboratory analyses, it is common practice to extract bitumen from source rock by washing the sample with an organic solvent. In field applications however, the organic solvents used in the laboratory (such as methylene chloride or chloroform) are not practical because they are expensive and toxic. Ongoing research at Schlumberger has led to the development of a green solvent that is environmentally and economically acceptable to be deployed at large scale in the subsurface.

The composition of bitumen varies based on factors such as source and maturity. Thus, the optimum formulation for bitumen removal also varies from region to region. The proposed green solvent for bitumen dissolution is a blend of three separate solvents. The effectiveness of removing bitumen is assessed on a ternary phase diagram, with the pure solvents at the apex of this diagram. Each experimental trial is represented as a point on the plot, with the color/size of the point representing the effectiveness of the blend in removing bitumen. A color scale Is used to represent solvent efficacy relative to methylene chloride.

Initial laboratory tests demonstrate that this green solvent extracts bitumen as efficiently as the toxic solvents and generates similar enhancement in the porosity of the source rock. Furthermore, these alternative solvents are readily available today, are inexpensive, and are environmentally permissible for subsurface applications.

577752 Co-Aggregation of Asphaltenes and Inorganic Solids in Diluted Bitumen

Athabasca bitumen contains over 10% asphaltenes and around 1% inorganic solids following the froth extraction process. Asphaltenes are a class of large aromatic organic compounds. A recent study found that the asphaltene precipitation rate increases significantly when inorganic solids were present in the mixture. [1] The results were modeled as the inorganic solids acting as collection sites for asphaltenes, which allowed them to be more easily separated by centrifugation. This new observation opens up new areas of study; specifically, determining the role of asphaltene adsorption onto solid surfaces during the dynamic process of phase separation. In this study, we aim to investigate the structure of asphaltenes and inorganic solid clusters that form during precipitation to determine the aggregation mechanism and dynamics. Athabasca bitumen was diluted with a heptane-toluene blend at low concentrations to promote asphaltene adsorption onto solid surfaces, but below the asphaltene bulk phase transition. After waiting different durations, additional heptane-toluene diluent was added to precipitate asphaltenes. The rate and quantity of solids collected was recorded. Additionally, the precipitated clusters were cut by focused ion beam (FIB) milling and images were taken by scanning electron microscopy (SEM). A large quantity of non-uniformly distributed ultra-fine solids (<200nm) were found in the precipitated asphaltenes clusters. A greater quantity of inorganic solids were detected in samples that were aged for a longer time at the lower diluent ratio. These results suggest that ultra-fine solids aggregate with asphaltenes and create nano-scale clusters (<500nm) that then further interact through cluster-cluster aggregation (>1 micron). Additional investigation of the solids properties that promote association with asphaltenes during precipitation is ongoing.

577553 Zero-Resistance Ammetry As a Low Power Approach to Monitor Microbially Influenced Corrosion

We are developing an electrochemical sensing method that uses zero-resistance ammetry (ZRA) measurements along a pipeline to detect microbially influenced corrosion (MIC). In order to develop ZRA as an MIC monitoring tool, we use split chamber ZRA (SC-ZRA) incubations that deploy two steel working electrodes (WE1 and WE2) in chambers separated a salt bridge. Subsequently, one chamber is inoculated with an individual or consortia of microorganisms, and ZRA and potential measurements can be made in conjunction with evaluations of microbiological activities. This approach allows us to separate the abiotic and biotic corrosion mechanisms. During monitoring, the direction and magnitude of the current produced can be indicative of the mechanisms and extents of MIC. Additionally, the SC-ZRA setup mimics the heterogenous biofilm coverage of metal surfaces that leads to MIC, and could be used to establish the electrochemical signatures of MIC. In this work, detection of MIC will be carried out under anoxic and flow conditions. We have obtained samples from several natural gas transmission pipelines and are conducting preliminary physicochemical and microbiological characterizations of the slurries.

581363 A New Approach for Gas Hydrates Transportability under Multiphase Flow

This contribution focus on the coupling of gas hydrate formation under multiphase flow encountered in oil and gas production. Our new approach considers the interaction between the hydrate porous structure and the oil continuous phase. Such interaction is herein called 'the sponge approach', where capillary filling-up with time is considered as the mass transfer limitation process. The filling-up process competes with water displacement from the sponge due to capillarity forces. In the case of high oil-water interfacial tensions or small wetted angles in the water-oil-hydrate interface, capillarity forces are higher and water tends to accumulate around the outer surface of the particle. The particle then becomes wet, enhancing agglomeration due to the possibility of crystalline bridge construction after the collision of two particles. The use of Anti-Agglomerants (A-A), which often exhibit surfactant activity, reduces both interfacial tension and wetted angles. Therefore, the kinetic model proves to be consistent in explaining the mechanisms of anti-agglomeration. This model gives new insight into the physical-chemical interactions that are coupled with shear forces under multiphase flow, resulting in a more comprehensive and robust engineering tool for hydrate management.

562614 Modelling Ice and Wax Deposition in a Pipeline in the Arctic Environment

In the cold regions, such as Arctic environment, fluid temperature in pipeline can drop below the freezing point of water. Water even in trace amounts may form ice and lead to pipeline blockages and associated risks. The restart of the Poplar pipeline system which gathers crude oil from Montana and North Dakota was delayed due to a pipeline blockage that was caused by ice plug. The concern of ice formation in the Trans-Alyeska Pipeline System also leads the company to evaluate the ice risks to the pipeline. Ice can deposit on pipe wall In addition to wax deposition. . Solid formation on pipeline surface can lead to flow assurance and process safety issues, such as blockage of pipeline, pipeline component failure, and the release of hazardous liquid. The blockage of pipeline can cause additional burden or failure to pumping system. Remediating the plugging requires shutdown of pipeline operation, which cause tremendous cost and delay to the entire production system. Ice and wax deposition in pipeline is a slow process. Pigging operation can be used to remove the deposits on pipeline surface. However, if deposition is too thick, pipeline blockage can still occur. In order to prevent pipeline blockage, ice and wax deposition rates are required to be estimated.

This paper investigates ice and wax deposition rates in a 1.5 km pipeline under various operating conditions, such as inlet temperature, ambient temperature, and fluid properties. A fundamental model for both ice and wax deposition is proposed using first principles of heat and mass transfer. The interaction between water droplets (or ice particles) and wax is analysed and compared with experimental data.

553418 Towards Field Wide Production Optimization for Upstream Field Recovery; A Holistic Analysis and Optimisation Framework of the Upstream Field from Reservoir to Host Facility.

Towards Field Wide Production Optimization For Upstream Field Recovery
Shakeel Ramjanee, CEng MEng, MIChemE (Imperial College London)
Supervisors: Dr Benoit Chachuat, Professor Ann Muggeridge & Professor Nilay Shah

Integrated Asset Modelling (also known as Integrated Production System Modelling - IPSM) is the modelling and simulation of an entire production facility consisting of both subsurface and surface elements. The intent is to holistically capture the complex interactions between the individual components of the system.

Historically asset modelling has entailed discrete reservoir, wells and facility models; however this mindset is slowly altering towards integrated asset modelling for production forecasts and optimisation whereby both system imposed constraints i.e. fluid flow from the reservoir and surface handling constraints can be considered simultaneously.

The full benefits of the integrated simulation tools employed within industry are not fully realised due to a number of limitations; complexity of the reservoir models and hence limited usability in day to day production optimisation, overcoming the 'silo' discipline mentality that is common within interdisciplinary surveillance teams, simplification of integrated models is often in tension with accuracy and lastly, the limited capability of current tools to apply multi objective optimization analysis.

With these challenges, it can often be difficult to justify the effort and investment required to develop and deploy an integrated asset model. Despite these practical challenges, integrated asset modelling, if executed correctly can provide one of the best means of optimising production throughout field life with direct realized benefits such as extended field life and sustained production uptime.

The main objective of this research study is to understand the opportunities and limitations with the current existing industry integrated packages and develop a viable, representative alternative that is applicable for day to day surveillance and medium to long term field production optimisation. This research also aims to formally outline the drivers for integrated asset modelling followed by a review of the most common integrated asset modelling tools on the market in addition to review of the optimisation and reservoir modelling simplification methods available in literature.

An IPSM model has been constructed and benchmarked via an integrated modelling tool employed within the upstream industry. This has then been evaluated against a derived proxy model via a HDMR (High Definition Meta-modelling Representation) general purpose meta modelling tool at Imperial College London for comparison including a global sensitivity analysis of the variables selected for the model development. This mathematical proxy model is then compared against the coupling of a full physics reservoir model to the integrated field model with telling comparisons, performance and crucially results. These models (which includes an aquifer driven case) are then utilised as inputs to further optimisation studies to evaluate the impact of short term decision making on the long term decision horizon i.e. field performance.

The ultimate aim is to develop a viable alternative which incorporates aspects of secondary recovery (water and/or gas injection) and allows for short/medium term horizon analysis for improved integrated field management and hence, improved production optimization modelling capability.

576305 Settling Characteristics of Weighting Material in the OIL-Based Drilling Fluids

Maintaining a uniform drilling fluid density along the downhole and controlling the downhole pressure are important tasks for successful and safe operation of the drilling process. Settling of the barite particles, which act as the weighting material, can lead to density variations in the downhole as high as 0.5 kg/L, and severe operational problems such as loss of control, lost circulation and stuck pipe, etc. Further, the local density of the drilling fluid or mud plays an important role in preventing gas entrainment and fluid formation in the downhole. Despite continuous efforts by the drilling industry, mitigation of the sagging of weighting materials in drilling fluids continues to be a challenge faced by the oil industry. Therefore, there is a need to understand the settling characteristics of barite particles in oil-based drilling muds. Though there are several publications reported on this topic, no publication has reported on the relation between barite sag and rheological properties of the drilling fluid. Hence in the present work, an attempt has been made to relate the static and dynamic sag potentials with the rheological properties of the oil-based drilling fluid.

Rheological properties of the drilling fluid have been measured using the ARES G2 tensiometer (TA instruments) with a cup-and-bob geometry. A barite sag phenomenon in oil-based drilling fluids was experimentally investigated under static and dynamic conditions. The settling characteristics of the barite particles have been quantified using pressure drop method and gamma densitometry, for the first time. Experiments were performed in a static cell (with no shear) and in a Taylor Couette cell (with low shear stress, to simulate the real operating conditions in the drilling process). Preliminary results indicated the shear thinning behavior of the drilling fluids with the existence of yield stresses at all temperatures (10 - 70°C), as evidenced by the decreasing slope of the stress profile with an increase in the shear rate (0.01 - 1000 S^{-1}). The temperature dependence was also clearly visible as the viscosity decreased exponentially with increasing temperature. Gamma densitometry experiments using a static cell revealed that the fluid density decreased at an axial location of 6 cm (near the top of the test section) with time. The decrease in the fluid density near the top of the test section occurred due to the gravitational settling of high-density barite particles. On the other hand, the fluid density at lower axial locations (1, 3 and 5 cm from the bottom) increased with time. Interestingly, the increment in the fluid density was approximately the same at all three axial locations, and could be attributed to the gel forming nature of the drilling fluid. The rate of change in the density could be correlated to the sedimentation rate in both static and Taylor Couette cells. Additionally, an attempt has also been made to relate the rate of barite particle settling to the rheological properties of the drilling fluid.

571792 Use of Eco-Friendly Amine Based Hydrate Inhibitors for Hydrate Mitigation and Offshore Flow Assurance

Gas hydrate is ice-like crystalline compounds that are formed under high pressure and low-temperature conditions in the subsea lines causing an unwanted blockages and safety threat to close by operators. Annually, the industry spends about 1 billion US $ on the prevention of gas hydrates and they are considered to be a major flow assurance concern for the oil and gas sector. In order to prevent the hydrate formation the industry heavily relies on the use of thermodynamic hydrate inhibitors such as methanol and mono-ethylene glycol (MEG). These inhibitors prevent hydrate formation by shifting hydrate vapor-liquid equilibrium to lower temperature values. However, the major issue with the use of these inhibitors is that they are required in large dosage (> 60 wt%), high cost, high toxicity, flammability and environmental constraints associated with their disposal. So the aim of this work is to present a new class of eco-friendly, low dosage and cost effective dual functional inhibitors to prevent hydrate formation in offshore subsea lines.

A series of experiments were conducted under different pressure conditions (40-100 bars) using rocking cell assembly (RC-5) and pure methane gas. The RC-5 mimics a subsea line and can be operated up to 200 bars with a temperature range of -10 ^{o}C to 60 ^{o}C. The inhibition performance of selected amine based eco-friendly hydrate inhibitors with and without synergists was evaluated at different dosages and pressure conditions. It was observed that the selected eco-friendly inhibitors can function as thermodynamic and kinetic inhibitors both simultaneously. The addition of synergist enhances their inhibition performance and help in reducing their required dosage. This work is beneficial for industry as it may help to reduce their operating cost, reduce their environmental foot print and assure smooth energy transfer.

About Authors:

Prof. Majeda Khraisheh and Dr. M Fahed Qureshi are active researchers in the area of flow assurance. Their work focuses primarily on applied research in the area gas hydrates considering the offshore industrial needs and addressing the environmental constraints associated with commercial hydrate inhibitors. They can be reached at m.khraisheh@qu.edu.qa and fahed.qureshi@yahoo.com .

571867 Investigating the SI Hydrate Crystal Formation at Different Produced Water Levels for Offshore Flow Assurance

As the oil fields mature, the amount of produced water increases. The excess water production is likely to lead to problems such as corrosion, gas hydrate formation, salt deposition, and water disposal issues. Gas hydrate is ice-like crystalline compounds that are formed under high pressure and low-temperature conditions in the sub-sea lines causing an unwanted blockages and safety threat to close by operators. Annually, the industry spends about 1 billion US \$ on the prevention of gas hydrates and they are considered to be a major flow assurance concern for the oil and gas sector. In order to prevent the hydrate formation the industry heavily relies on the use of chemical inhibitors such as thermodynamic hydrate inhibitors and kinetic hydrate inhibitors. These inhibitors are injected into the pipeline and they function by shifting hydrate vapor-liquid equilibrium to lower temperature values or by delaying the hydrate formation time. The amount of water present in the subsea lines is likely to affect the performance of these chemical inhibitors and depending on the level of water present the dosage of the inhibitor may need to be altered. So the aim of this project is to investigate the hydrate SI crystal formation at different water levels and see the effect of different water levels on the performance of selected gas hydrate inhibitors.

A series of experiments were conducted under different pressure conditions (40-100 bars) using rocking cell assembly (RC-5) and pure methane gas. The RC-5 mimics a subsea line and can be operated up to 200 bars with a temperature range of -10 ^{o}C to 60 ^{o}C. The inhibition performance of selected eco-friendly hydrate inhibitors was evaluated at different water levels and the amount of hydrate formation that occurs at each water level was estimated. According to the preliminary experimental results, the different water levels do have an effect on the amount of hydrates produced and the inhibition performance of selected hydrate inhibitors. This work is highly beneficial for the oil and gas sector as the hydrate formation at different water levels has been examined along with the performance evaluation of the hydrate inhibitors at different water levels. This will help the industry to optimize their inhibitors dosages, reduce their operating cost and assure smooth energy transfer.

About Authors:

Prof. Majeda Khraisheh and Dr. Fahed Qureshi are active researchers in the area of flow assurance. Their work focuses primarily on applied research in the area gas hydrates considering the offshore industrial needs and addressing the environmental constraints associated with commercial hydrate inhibitors. They can be reached at m.khraisheh@qu.edu.qa and fahed.qureshi@yahoo.com .

577982 Kinetic Hydrate Inhibitors: Gas Hydrate Induction Time Probability Distributions and Deviations from Classical Nucleation Theory

As subsea oil and gas production moves toward deeper water, high-pressure and low-temperature conditions leads to increased risk of gas hydrate formation. The agglomeration and aggregation of hydrate particles can then lead to costly pipeline plugging events. Various inhibitor chemicals can be injected into flowlines to prevent or delay hydrate formation. Unlike Thermodynamic Hydrate Inhibitors (THIs) such as methanol however, Kinetic Hydrate Inhibitors (KHIs) do not affect the hydrate phase equilibrium curve, rather they delay hydrate formation. Robust KHI performance qualification thus requires formation itself to be measured rather than the dissociation curve (which can be used to qualify THI performance). This poses a challenge since hydrate nucleation is stochastic and thus necessitates a high number of independent measurements of hydrate formation events, a requirement which is challenging to realise with conventional apparatus that are only capable of relatviely slow temperature ramping.

Here we detail the use of a high-pressure-stirred-automated-lag-time-apparatus (HPS-ALTA) to enable efficient quantification of methane hydrate formation probability in the presence of a KHI. Typically 100 independent hydrate formation events are obtained for each experimental condition, enabling the construction of smooth induction time probability distributions. We confirm previous findings in that we demonstrate that increasing KHI concentration leads to increased induction times, albeit with diminishing returns as the concentration is increased in the (1-2) wt% range. We also show that although induction times distributions are exponentially distributed for the KHI-free system (consistent with predictions from Classical Nucleation Theory), this is not the case in the presence of a KHI where our experimental data can be best described by a gamma distribution with lower relative standard deviation. This result has significance not only for understanding hydrate formation risk in the presence of KHIs but also for improving our theoretical understanding of the mechanism by which KHIs delay hydrate formation.

571307 Structure H Hydrate Formation with Lmgs and CO_2, in Ice and Water

Structure H hydrate formation with LMGS and CO_2, in Ice and Water

Namrata Gaikwad[1], Omkar S. Kushwaha[1], Praveen Linga[3], Jitendra Sangwai[2], Rajnish Kumar[1]*

[1] Department of Chemical Engineering, Indian Institute of Technology, Chennai, 600036, India

[2] Department of Ocean Engineering, Petroleum Program, Indian Institute of Technology, Chennai, 600036, India

[3] Department of Chemical and Biomolecular Engineering, National University of Singapore, 117585, Singapore

Corresponding author: rajnish@iitm.ac.in

Keywords: Gas hydrates, CO_2 Capture, CH_4 Storage, Hydrate Kinetics, Hydrate Dissociation.

Abstract

Van der Waals radius of carbon dioxide molecules are significantly larger than methane and thus it is a reluctant help gas for sH hydrate formation. A successful attempt has been made to form structure H (sH) hydrate at low pressure from four different types of experiments in ice and water by employing methane and carbon-dioxide as a small molecule guest substance (SMGS) and polar tert butyl methyl ether (TBME) and hydrophobic cyclo-octane as a large molecule guest substance (LMGS). One of the ideas was to use different LMGS such that the effect of oxygen molecule containing LMGS which are slightly polar could be studied with those LMGS which are completely non-polar. Type-1 experiments were carried out to study the kinetics of sH hydrate formation and dissociation for methane as SMGS for two identified sH formers at 274K. Type-2 experiments were performed to study the kinetics of carbon dioxide sH hydrate formation and dissociation at 274K i.e, sH hydrate formation with liquid water, such data are very rare in literature! Type-3 experiments were studied to understand the formation and dissociation kinetics for carbon dioxide sH hydrate at 266K, the literature suggests that sH hydrate formation with CO_2 and LMGS is possible only at sub-zero temperature, i.e., with solid water. Type-4 experiments were done to study the effect of driving force on carbon dioxide sH hydrates formation in ice at 266K. Experimental data on avg. gas uptake, water to hydrate conversion, kinetics of methane and carbon dioxide sH hydrate formation and dissociation at low pressure has been obtained. This study shows that kinetics of gas uptake and water to hydrate conversion depends on the type of LMGS and driving force used. Experiments with

TBME as a LMGS have shown higher gas uptake and rapid hydrate formation for methane. However, the rate of hydrate growth for CO_2 hydrate was faster with TBME in ice and cyclo-octane in water. It is observed that CO_2 molecules interact strongly with TBME in the liquid phase which might be acting as an inhibitor in the formation of hydrate. The system containing TBME as a LMGS has displayed a higher dissociation rate compared to cyclo-octane for both methane and carbon-dioxide gas.

575481 Understanding the Influence of Thermodynamic Hydrate Inhibitors on the Mechanics of Cyclopentane Hydrate Films Using Interfacial Rheology with Subphase Exchange

Clathrate or gas hydrates are known to form at water-hydrocarbon and water-natural gas interfaces, generally at low temperatures and high pressures, to trap the guest hydrocarbon or gas molecules in crystalline cages of ice-like water. Gas hydrates are of interest to the oil and gas recovery community as rapid nucleation and growth of these solids can block flow lines and create significant problems. Thus, extensive efforts have been made to link hydrate film nucleation, growth, and dissociation using bulk hydrate formation and measurement techniques. However, the link between film nucleation, growth, and dissociation remains unclear because of the difficulties associated with generalizing bulk results to processes comprised of interfacial phenomena. Recently, interfacial rheological methods have been employed to investigate the growth kinetics and film mechanics of cyclopentane hydrate films that conveniently form at atmospheric pressure. By modifying our interfacial rheology setup to include the ability to exchange subphase fluid without disturbing the interface, we are able to measure hydrate film mechanical properties before, during, and after the introduction of thermodynamic hydrate inhibitors to gain fundamental insight into the interfacial processes associated with film nucleation, growth, and dissociation. We will present our results on the formation and dissociation of cyclopentane hydrate films in the presence of different thermodynamic inhibitors and discuss how this might influence bulk rheological properties.

570674 Insight into the Hydrate Nucleation of Structure II Propane Hydrates from Molecular Simulation

Among the three main types of gas hydrate crystal structure, structure II (sII) hydrates are the most prevalent in oil and gas pipe line as propane and butane are commonly present, and these these molecules can only fit into the large cage of sII. While the nucleation of structure I hydrates has been been studied by a number of groups, there is still no insight into how sII hydrates forms, in particular when only a large guest, such as propane, is present to fill the large cages while leaving all the small cages empty. We performed molecular dynamic simulations for the nucleation of propane hydrates to capture the incipient cages of sII hydrates and to investigate the nucleation mechanism and crystallization process of sII hydrates. Results will be shown detailing the cages formed during the nucleation and the ordering leading to the crystalline structure.

559776 Modeling the Effects of Interfaces on the Nucleation of Gas Hydrates and Ice in Porous Media

Gas hydrates are water crystalline structures stabilized by the presence of gas molecules. Gas hydrates occur naturally in the permafrost or in ocean-floor sediments at low temperatures and high pressures. Predictions on the amount of methane trapped within gas hydrates contained in marine sediments amounts to three orders of magnitude the worldwide conventional natural-gas reserves. Gas hydrates are an untapped energy resource and a potential contributor to global warming. Mathematical modeling of the stability, formation and dissociation of gas hydrates within sediments is a necessary tool for their management. Due to their similarity to ice, aspects of the behavior of gas hydrates can be modeled similarly to the treatment of ice.

Heterogeneous nucleation takes place during the formation of gas-hydrates and ice in porous media. The presence of surfaces and interfaces within porous media can effectively lower the energy barrier for nucleation. The present work will introduce mathematical models developed in order to describe the effects on energy barriers during heterogeneous nucleation of ice or gas hydrates in porous media. Three different types of interfaces found in these environments will be discussed: a) gas/liquid interfaces, b) solid surfaces, and c) gas/liquid/solid triple boundary lines.

Model predictions suggest that nucleation is more energetically favorable on solid surfaces than at the gas/liquid interface or triple boundary line. In the case of solid surfaces, nucleation is favored within concave surfaces (e.g., inside porous spaces) and to a lesser extent on convex surfaces (e.g., particulates), while flat surfaces act as the behavioral boundary between these two cases. Additionally, hydrophilicity of solid surfaces enhances the positive effect of these surfaces towards nucleation. In the case of gas/liquid interfaces and triple boundary lines, nucleation and crystal growth occur preferentially within the liquid phase rather on the gas phase, and it is favored when the gas/liquid interfaces are present between slightly hydrophobic surfaces.

570727 A Fundamental Comparison of the Stability and Dissociation Aspects of Structure I and Structure II Hydrates from the Perspective of Feasible Natural Gas Storage

The hydrate based approach to storing natural gas (known as Solidified Natural Gas (SNG) technology) has been gaining popularity in recent times owing to its environmentally benign operation, large volumetric capacity and non-explosive character. By transitioning from sI to sII hydrates through the introduction of thermodynamic promoters like tetrahydrofuran (THF) to the system, researchers have not only been able to favourably tune the thermodynamic operating conditions for hydrate formation (significantly higher temperature and lower pressure for hydrate formation), but also demonstrated remarkable kinetic performance during hydrate formation at the revised operating conditions. However, the true feasibility of this approach lies in the long term stability and thus storability of mixed CH_4-THF sII hydrates at atmospheric pressure and how it compares with that for pure CH_4 sI hydrate under similar storage conditions.

According to thermodynamic data, the equilibrium temperature for mixed CH_4-THF hydrates (considering 6 mol % of THF in the system) at 0.12 MPa is 277.7K, while the same for pure CH_4 hydrate at 0.1 MPa is 193 K. Thus ideally, for storage at atmospheric pressure and moderately low temperatures (268-271 K), mixed CH_4-THF sII hydrate should be able to demonstrate complete stability while the pure CH_4 sI counterpart should dissociate rather rapidly. In the current study - a first of its kind, a like for like comparison was made for the stability and dissociation characteristics of methodologically identical pure CH_4 (sI) and mixed CH_4-THF (sII) hydrates. Cylindrical sI and sII hydrate pellets were synthesized using a one of a kind SNG technology prototype. Synthesized hydrate pellets were transferred to storage units for seven to ten day stability tests at atmospheric pressure and moderately low temperatures (~268-271 K). Rate of hydrate dissociation via thermal stimulation using room temperature water stream was separately studied for the synthesised sI and sII pellets. Results indicate that while sII pellets were completely stable at the investigated storage conditions, sI pellets started dissociating as soon as the tests were initiated. This demonstrates clearly the remarkable stability of sII hydrates when stored at atmospheric pressure and moderately low temperatures, thus highlighting a major downstream advantage of opting for the mixed sII hydrate route as far as SNG technology is concerned. The results obtained from this study are of great importance from both technical and economic standpoints and should go a long way in ensuring SNG technology as an attractive alternative for mid to long term storage of natural gas.

570540 Container-Free Gas Hydrate Nucleation Using High-Pressure Acoustic Levitation

Understanding hydrate nucleation is of direct relevance to the field of flow assurance in the oil and gas industry. However, due to its stochastic nature and potential apparatus dependencies, hydrate nucleation can be difficult to generalise. Conventionally, studies are performed with high-pressure vessels wherein solid-liquid interfaces generate potential hydrate nucleation sites that are non-trivial to characterise (e.g. microscopic imperfections). To overcome such constraints, we demonstrate here the first statistically significant set of truly container-free hydrate formation measurements using a high-pressure acoustic levitator. In our experiments, water droplets with well-defined geometries were acoustically levitated within a natural gas atmosphere at approximately 65 bar and 7 °C. Hydrate formation on the interface of the ellipsoidal droplets was directly imaged and used to extract induction times. Cumulative probability distribution functions for these induction times at a controlled sub-cooling were constructed and compared with those obtained from conventional measurements and with predictions from classical nucleation theory. Hydrate dissociation within the acoustic field was also compared to that measured in a high-pressure autoclave.

AUTHOR INDEX

AUTHOR INDEX